植物乳杆菌的益生功能及其作用机制

李川 曹君 著

辽宁大学出版社
Liaoning University Press

图书在版编目（CIP）数据

植物乳杆菌的益生功能及其作用机制/李川，曹君著.—沈阳：辽宁大学出版社，2018.7
ISBN 978-7-5610-9193-7

Ⅰ.①植… Ⅱ.①李…②曹… Ⅲ.①乳酸细菌—研究 Ⅳ.①Q939.11

中国版本图书馆 CIP 数据核字（2018）第 088486 号

植物乳杆菌的益生功能及其作用机制
ZHIWU RUGANJUN DE YISHENG GONGNENG JI QI ZUOYONG JIZHI

出 版 者：	辽宁大学出版社有限责任公司
	（地址：沈阳市皇姑区崇山中路66号　邮政编码：110036）
印 刷 者：	沈阳海世达印务有限公司
发 行 者：	辽宁大学出版社有限责任公司
幅面尺寸：	170mm×240mm
印　　张：	10.75
字　　数：	170千字
出版时间：	2019年4月第1版
印刷时间：	2019年4月第1次印刷
责任编辑：	于盈盈
封面设计：	优盛文化
责任校对：	齐　阅

书　　号： ISBN 978-7-5610-9193-7
定　　价： 39.00元

联系电话：024-86864613
邮购热线：024-86830665
网　　址：http://press.lnu.edu.cn
电子邮件：lnupress@vip.163.com

前　言

　　益生菌是由单一或多种微生物构成的活菌，当摄入一定剂量时，能通过改善宿主肠道微生态平衡来促进人体健康。益生菌具有促进肠道菌群平衡、缓解代谢综合征和免疫调节等多种功能。本书作以本实验室前期分离的体外活性较强的植物乳杆菌 NCU116 为研究对象，探讨其对肠道菌群的调节作用；对便秘、高脂血症、脂肪肝、结肠炎和糖尿病的缓解作用；并基于代谢组学技术检测其对高脂血症和糖尿病血清中的特征代谢标志物的影响，主要研究内容与结论归纳如下：

　　（1）小鼠被随机分为 4 组，给予小鼠灌胃不同剂量植物乳杆菌 NCU116 或生理盐水，连续 5 周。与正常组相比，植物乳杆菌 NCU116 组小鼠肠道内乳酸杆菌、双歧杆菌的菌群显著升高。此外，该菌还在产生短链脂肪酸、抑制氧化应激和调节血清细胞因子方面具有一定的作用。

　　（2）通过洛哌丁胺皮下注射建立小鼠便秘模型，灌胃不同剂量植物乳杆菌 NCU116，连续 15 天。与便秘模型组相比，植物乳杆菌 NCU116 组在粪便指标、肠道推进率、短链脂肪酸、结肠病理与结肠间质细胞免疫表达方面具有显著改善。结果证实，植物乳杆菌 NCU116 对洛哌丁胺诱导的小鼠便秘症状具有缓解作用。

　　（3）研究植物乳杆菌 NCU116 对高脂高胆固醇饲料喂养大鼠胆固醇的降低作用。大鼠随机分为 4 组：正常组，高脂模型组，高脂模型＋植物乳杆菌 NCU116 低、高剂量组，连续灌胃干预 5 周。结果显示，植物乳杆菌 NCU116 表现出降低血脂水平、修复胰腺和脂肪组织损伤的能力。此外，该益生菌还可显著提高低密度脂蛋白受体（LDL receptor）和胆固醇 7α- 羟化酶（CYP7A1）等基因的表达。表明植物乳杆菌 NCU116 可能是通过调节低密度脂蛋白受体和胆固醇 7α- 羟化酶的基因表达进而调控脂质代谢和降低胆固醇水平。

　　（4）研究植物乳杆菌 NCU116 对高脂饮食诱导非酒精性脂肪肝大鼠肝功能、氧化应激和脂质代谢的影响。实验 5 周后，植物乳杆菌 NCU116 组表现出修复肝功能、缓解肝病理损伤、降低肝脏脂肪堆积的功能。此外，该菌还能显著降低血清内毒素和促炎因子、调节肠道菌群、调节脂质代谢基因表达。结果显示，植物乳杆菌 NCU116 缓解非酒精性脂肪肝可能是通过抑制脂肪合成、促进脂肪分解与氧化等相关基因表达来实现的。

（5）建立基于超高效液相色谱串联四级杆飞行时间质谱（UPLC-Q-TOF/MS）的代谢组学方法，分析植物乳杆菌 NCU116 对高脂饮食大鼠血清中代谢标记物的影响。通过非配对 t 检验、主成分分析、偏最小二乘法分析和分层聚类分析方法可知，4 组样本具有较好的分离度。高脂饮食对大鼠血清中亚精胺、维生素 B5、吲哚丙烯酸、吲哚、5-羟吲哚乙醛、甘氨胆酸、胆氯素、羟基乙酸、亮氨酸、2-苯基乙醇葡萄糖苷酸、牛磺胆酸、2-花生酰基甘油、孕二醇-3-葡萄糖苷酸等化合物代谢均有影响。但植物乳杆菌 NCU116 仅在吲哚丙烯酸、甘氨胆酸、胆氯素、羟基乙酸、牛磺胆酸、2-花生酰基甘油和孕二醇-3-葡萄糖苷酸等生物标记物上发挥较好的调节作用。提示该菌可通过这些生物标记物来调节机体内脂质、葡萄糖、氨基酸等代谢途径发挥抑制高脂血症作用。

（6）采用三硝基苯磺酸（TNBS）诱导建立结肠炎小鼠模型，研究植物乳杆菌 NCU116 及其发酵胡萝卜汁对该症状的影响。小鼠被随机分正常组、结肠炎模型组、植物乳杆菌 NCU116 组、植物乳杆菌 NCU116 发酵胡萝卜汁组和未发酵胡萝卜汁组，连续灌胃 5 周。结果显示，经植物乳杆菌 NCU116 及其发酵胡萝卜汁干预后，结肠炎小鼠体重较快恢复并有所增长，氧化应激和促炎因子水平降低，结肠粪便短链脂肪酸水平显著升高，结肠黏膜损伤明显降低。通过与未发酵胡萝卜汁对比可知，植物乳杆菌 NCU116 及其发酵胡萝卜汁对结肠炎的缓解作用与植物乳杆菌 NCU116 的供给密切相关。

（7）探讨植物乳杆菌 NCU116 及其发酵胡萝卜汁对高脂饮食结合链脲佐菌素诱导的 II 型糖尿病大鼠降血糖作用及其机制。大鼠被随机分为正常组、糖尿病组、植物乳杆菌 NCU116 组、植物乳杆菌 NCU116 发酵胡萝卜汁组和未发酵胡萝卜汁组，连续灌胃 5 周。结果显示，植物乳杆菌 NCU116 及其发酵胡萝卜汁对大鼠血糖、血脂、激素具有调节作用，同时对短链脂肪酸水平具有升高作用，还能修复氧化应激损伤、胰腺与肾脏病理损伤，并能调节低密度脂蛋白受体、胆固醇 7α-羟化酶、葡萄糖转运蛋白 4、过氧化物酶体增殖因子活化受体 α 和 γ 的基因表达。本实验首次证实植物乳杆菌 NCU116 及其发酵胡萝卜汁对 II 型糖尿病具有一定的缓解作用。

（8）利用 UPLC-Q-TOF/MS 等代谢组学技术来探讨植物乳杆菌 NCU116 及其发酵胡萝卜汁对 II 型糖尿病大鼠代谢标记物变化的影响。通过非配对 t 检验，主成分分析和偏最小二乘法分析可知，植物乳杆菌 NCU116 及其发酵胡萝卜汁对糖尿病大鼠血清中腺苷、5-羟吲哚乙醛、脯氨酸、甘氨胆酸、牛磺鹅去氧胆酸、鞘氨醇、茶碱和牛磺胆酸代谢均有影响，并通过调节机体内葡萄糖、脂肪酸、胆酸和氨基酸等代谢途径发挥抗糖尿病作用。另外，植物乳杆菌 NCU116 及其发酵胡萝卜汁表现出的缓解糖尿病症状的功能与植物乳杆菌 NCU116 的补充有关。

PREFACE

Probiotics have been defined as "live microorganisms which, when administered in adequate amounts, confer a health benefit on the host". Probiotics have several kinds of functions, such as regulating intestinal flora, alleviating metabolic syndrome and immunomodulatory. *Lactobacillus plantarum* NCU116 was recently isolated from pickled vegetables. Previously reports have showed that this bacterium is characterized with good performance *in vitro*. However, the character of the probiotic *in vivo* is unclear. The present study was to investigate the effects of *L. plantarum* NCU116 on intestinal flora, constipation, hyperlipidaemia, non-alcoholic fatty liver disease, inflammatory bowel disease, diabetes; and serum metabolomics methods were developed to investigate the effect of metabolites of *L. plantarum* NCU116 on hyperlipidaemia and diabetes. The main conclusions obtained in this dissertation are summarized as follows:

(1) Mice were randomly divided into four groups and orally administrated saline and three doses of *L. plantarum* NCU116 groups (NCU116-L, 10^7 CFU/mL; NCU116-M, 10^8 CFU/mL; NCU116-H, 10^9 CFU/mL; respectively) for five weeks. Compared with the normal group, *Lactobacillus*, *Bifidobacterium*, short chain fatty acids were increased in the groups that received *L. plantarum* NCU116. In addition, the probiotic reduced the oxidative stress and proinflammatory factor in serum.

(2) This part of study examined the effects of *L. plantarum* NCU116 on loperamide-induced constipation in a mouse model. Loperamide was injected subcutaneously to induce constipation. Animals were divided to five groups: normal group, constipation group, constipation plus three doses of *L. plantarum* NCU116 groups (NCU116-L, 10^7 CFU/mL; NCU116-M, 10^8 CFU/mL; NCU116-H, 10^9 CFU/mL; respectively). Mice were treated with the probiotic for 15 days to assess the anti-constipation effects. Fecal parameters,

intestinal transit ratio and the production of fecal short chain fatty acids, histological of colon and immunohistochemical in colonic cells of Cajal (ICC) by *c-kit* were all improved in *L. plantarum* NCU116 treated mice as compared to the constipation group. These results demonstrated that *L. plantarum* NCU116 enhanced gastrointestinal transit and alleviated loperamide-induced constipation.

(3) The cholesterol-lowering effect of *L. plantarum* NCU116 on lipid metabolism of rats fed on a high fat diet was investigated. Sprague-Dawley rats were randomly divided into normal diet (ND) group, high fat diet (HFD) group, HFD plus *L. plantarum* NCU116 groups with two different doses (NCU116-L, 10^8 CFU/mL; NCU116-H, 10^9 CFU/mL). After treatment for 5 weeks, *L. plantarum* NCU116 had the potential ability to regulate lipid metabolism levels, morphology of pancreas and adipose tissues. In addition, the bacterium significantly improved gene expression of low-density lipoprotein (LDL) receptor and cholesterol 7α-hydroxylase (CYP7A1). These results suggested that *L. plantarum* NCU116 was able to alter lipid metabolism and reduce the cholesterol level, in particular, in the rats on a high fat diet through regulating gene expression of key factors relating to LDL receptor and CYP7A1.

(4) The effects of *L. plantarum* NCU116 on liver function, oxidative stress and lipid metabolism in rats with high fat diet induced non-alcoholic fatty liver disease (NAFLD) were studied. Treatment of *L. plantarum* NCU116 for 5 weeks was found to restore liver function, liver morphology and oxidative stress in rats with NAFLD, and decrease the levels of fat accumulation in liver. In addition, the bacterium significantly reduced endotoxin and proinflammatory cytokines, and regulated bacterial flora in the colon and the expression of lipid metabolism in the liver. These results suggested that downregulating lipogenesis and upregulating genes expression related to lipolysis and fatty acid oxidation were involved in the beneficial effect of *L. plantarum* NCU116 on NAFLD.

(5) A metabolomics method based on ultra performance liquid chromatography quadrupole time-of-flight mass spectrometry (UPLC-Q-TOF/

MS) was developed to investigate the metabolites of *L. plantarum* NCU116 on serum from high fat fed rats. With *t* test unpaired analysis, a good separation of PCA, PLS-DA and HCA of four groups was achieved. 13 potential biomarkers, including spermidine, pantothenic acid, indoleacrylic acid, indole, 5-hydroxy indole acetaldehyde, glycocholic acid, biliverdin IX, glycolic acid, L-leucine, 2-phenylethanol glucuronide, taurocholic acid, 2-arachidonoylglycerol, pregnanediol-3-glucuronide have been identified in serum samples from high fat fed rats. In additoion supplement of *L. plantarum* NCU116 regulated the levels of indoleacrylic acid, glycocholic acid, glycolic acid, taurocholic acid, 2-arachidonoylglycerol, pregnanediol-3-glucuronide, and pathways of lipids, glucose and lipoprotein to alleviate hyperlipidaemia.

(6) Effect of carrot juice fermented with *L. plantarum* NCU116 on 2, 4, 6-trinitrobenzenesulfonic acid-induced inflammatory bowel disease (IBD) in mice was studied. Mice were randomly divided into five groups: normal, IBD, IBD plus *L. plantarum* NCU116 (NCU), IBD plus fermented carrot juice with *L. plantarum* NCU116 (FCJ) and IBD plus non-fermented carrot juice (NFCJ). Treatments of NCU and FCJ for 5 weeks were found to favorably regulate oxidative stress, proinflammatory factor, histological of colon; and increase body weight, short chain fatty acids compared with the IBD group. In addition, compared with the NFCJ group, the function of NCU and FCJ groups closely related to the supplement of *L. plantarum* NCU116.

(7) Effect of carrot juice fermented with *L. plantarum* NCU116 on high fat and low-dose streptozotocin (STZ)-induced type 2 diabetes in rats was studied. Rats were randomly divided into five groups: non-diabetes mellitus (NDM), un-treated diabetes mellitus (DM), DM plus *L. plantarum* NCU116 (NCU), DM plus fermented carrot juice with *L. plantarum* NCU116 (FCJ) and DM plus non-fermented carrot juice (NFCJ). Treatment of NCU and FCJ for 5 weeks were found to favorably regulate blood glucose, hormones and lipid metabolism in the diabetic rats, accompanied by an increase in short chain fatty acids (SCFA) in colon. In addition, NCU and FCJ restored the antioxidant capacity, morphology of pancreas and kidney, and up regulated mRNA of low-

density lipoprotein (LDL) receptor, cholesterol 7α-hydroxylase (CYP7A1), glucose transporter-4 (GLUT-4), peroxisome proliferator-activated receptor-α (PPAR-α), and peroxisome proliferator-activated receptor-γ (PPAR-γ). These results for the first time demonstrated that *L. plantarum* NCU116 and the fermented carrot juice had the potential ability to ameliorate type 2 diabetes in rats.

(8) A novel analysis method based on UPLC-Q-TOF/MS was developed to investigate the effects of *L. plantarum* NCU116 and its fermented carrot juice on the metabolites of sera from type 2 diabetic rats. With *t* test unpaired analysis, a good separation of PCA and PLS-DA of five groups was achieved. Eight potential biomarkers were identified in serum samples from type 2 diabetic rats. Biological pathways and processes were significantly changed by NCU and FCJ treatments. In addition, the NCU and FCJ groups showed good clustering than other groups, thus, the possible mechanism by which NCU and FCJ ameliorate type 2 diabetes in rats may be related to the action of *L. plantarum* NCU116 in the supplement.

目 录

第1章 引 言 / 001

    1.1 益生菌的概念 / 001

    1.2 乳酸杆菌的生物学特性 / 001

    1.3 乳酸杆菌的主要代谢产物 / 002

    1.4 乳酸杆菌的益生功能 / 004

    1.5 乳酸杆菌的安全性 / 008

    1.6 前景与挑战 / 008

    1.7 植物乳杆菌 / 009

    1.8 发酵胡萝卜汁 / 009

    1.9 代谢组学 / 010

    1.10 本书研究的主要内容和创新点 / 010

第2章 植物乳杆菌 NCU116 对小鼠肠道菌群的调节作用 / 013

    2.1 引 言 / 013

    2.2 实验部分 / 014

    2.3 结果与分析 / 017

    2.4 讨 论 / 024

    2.5 本章小结 / 025

第3章 植物乳杆菌 NCU116 对洛哌丁胺诱导小鼠便秘的缓解作用 / 026

    3.1 引 言 / 026

    3.2 实验部分 / 027

    3.3 结果与分析 / 030

    3.4 讨 论 / 036

    3.5 本章小结 / 038

第 4 章　植物乳杆菌 NCU116 对高脂饮食大鼠抑制血脂紊乱的机制　/ 039

  4.1　引　言　/ 039
  4.2　实验部分　/ 039
  4.3　结果与分析　/ 042
  4.4　讨　论　/ 050
  4.5　本章小结　/ 052

第 5 章　植物乳杆菌 NCU116 对大鼠脂肪肝病变的抑制作用　/ 053

  5.1　引　言　/ 053
  5.2　实验部分　/ 054
  5.3　结果与分析　/ 057
  5.4　讨　论　/ 066
  5.5　本章小结　/ 068

第 6 章　植物乳杆菌 NCU116 对高脂饮食大鼠血清代谢组学的影响　/ 069

  6.1　引　言　/ 069
  6.2　实验部分　/ 070
  6.3　结果与分析　/ 072
  6.4　讨　论　/ 078
  6.5　本章小结　/ 086

第 7 章　植物乳杆菌 NCU116 及发酵胡萝卜汁对小鼠结肠炎的缓解作用　/ 087

  7.1　引　言　/ 087
  7.2　实验部分　/ 088
  7.3　结果与分析　/ 090
  7.4　讨　论　/ 098
  7.5　本章小结　/ 099

第 8 章　植物乳杆菌 NCU116 及发酵胡萝卜汁对糖尿病大鼠的血糖改善机制研究　/　100

　　8.1　引　言　/　100

　　8.2　实验部分　/　101

　　8.3　结果与分析　/　103

　　8.4　讨　论　/　112

　　8.5　本章小结　/　116

第 9 章　植物乳杆菌 NCU116 及发酵胡萝卜汁对糖尿病大鼠的血清代谢组学初探　/　117

　　9.1　引　言　/　117

　　9.2　实验部分　/　118

　　9.3　结果与分析　/　119

　　9.4　讨　论　/　124

　　9.5　本章小结　/　128

第 10 章　结论与展望　/　129

　　10.1　本研究的主要结论　/　129

　　10.2　进一步的研究方向　/　131

参考文献　/　132

英文缩略语　/　156

第1章 引　言

1.1　益生菌的概念

益生菌的概念最早在 1965 年由 Lilly 等人提出,他们将其定义为"能产生促进生长因子的微生物"。[1]1989 年,Fuller 将益生菌的概念发展为"能够通过改善肠道菌群平衡来对宿主产生有益作用的活性微生物制剂"。[2]2001 年,联合国粮农组织和世界卫生组织(FAO/WHO)将益生菌定义为"由单一或多种微生物组成的活菌,当摄入一定剂量时,能通过改善宿主肠道微生态平衡来促进人体健康"。[3]

乳酸菌在发酵食品和乳制品中有着长久的安全使用历史,因而摄入乳酸菌获得的感染风险较低。食品中的乳酸菌大多来自哺乳动物的正常菌群,其安全性在众多的报道中得到证实。[4] 为了有效降低对抗生素的依赖,充分发挥益生菌在临床治疗上的作用,已受到广泛重视。

2010 年,卫生部印发了《可用于食品的菌种名单》,规定可食用益生菌为 21 种,其中 14 种是乳酸杆菌。因此,探讨乳酸杆菌对代谢活性和生理功能具有重要的意义。作为对人体和动物有益的微生物,乳酸杆菌具有多方面的功能:① 促进动物生长,提高食物消化率;[5] ② 黏附并定植于肠道;[6,7] ③ 拮抗肠道病原微生物生长,[8] 调节胃肠道菌群、维持肠道微生态平衡;[9] ④ 提高机体免疫力;[10] ⑤ 有效降低血清胆固醇含量,[11] 减少代谢综合征发病风险。[12]

1.2　乳酸杆菌的生物学特性

乳酸菌(Lactic acid bacteria, LAB)是由一类具有特定的形态学、新陈代谢、生理特征的革兰氏阳性细菌构成，主要包括乳杆菌属（*Lactobacillus*）、明串珠菌属（*Leuconostoc*）、片球菌属（*Pediococcus*）、链球菌属（*Streptococcus*）。[13]

乳酸杆菌属是乳酸菌中最大的一个属，[14]该属主要由革兰氏染色阳性、不产生芽孢、基因G+C含量通常小于50%、发酵糖类的主要终产物为乳酸的杆菌或球菌构成。[15, 16]乳酸杆菌属在发酵类型上可分为同型发酵（Homofermentative）、兼性异型发酵（Facultatively heterofermentative）、专性异型发酵（Obligately heterofermentative）。[17]这些发酵类型区分的生理学基础是细胞内是否存在糖类同型或异型发酵的关键酶（1, 6-二磷酸果糖醛缩酶和磷酸酮糖酶）。[18]目前，对乳酸杆菌内不同种进行区分在经典方法（发酵模式、乳酸构型、精氨酸水解、生长需求等）基础上发展了肽聚糖分析、乳酸脱氢酶电泳迁移率、DNA的G+C mol%和DNA杂交等更为准确快捷的方式。[19]

1.3 乳酸杆菌的主要代谢产物

1.3.1 乳酸和短链脂肪酸

乳酸杆菌发酵糖类主要产生乳酸和少量乙酸等短链脂肪酸。乳酸是一种重要的抑菌化合物。短链脂肪酸（Short chain fatty acids，SCFA）是指C1-C6的一元羧酸，这些水溶性羧酸易于被人体吸收，并产生多种生理功能。短链脂肪酸在体内由结肠细菌厌氧发酵多糖、寡糖等碳水化合物产生。[20-22]

短链脂肪酸对宿主有着重要的生理功能，乳酸和乙酸可以降低pH值，酸性环境有利于增强其他酸的生理活性；乳酸可以通过抑制一些致病菌的呼吸酶系的活力和对氨基酸的竞争吸收来达到抑菌目的；乙酸对真菌和部分致病菌的生长和繁殖具有抑制作用，丁酸在为肠上皮细胞提供营养方面具有重要意义。[23, 24]

1.3.2 共轭亚油酸

共轭亚油酸（Conjugated linoleic acids, CLA）是一种必需脂肪酸（不

能在人和动物体内合成），是亚油酸（十八碳二烯酸）的同分异构体。主要由 cis-9,trans-11 共轭亚油酸（c9,t11-CLA）和 trans-10,cis-12 CLA（t10,c12-CLA）组成。[25] 文献显示，共轭亚油酸具有抗肿瘤、降血糖、抗氧化及免疫调节等多种活性。[26] 乳酸杆菌为兼性厌氧菌，具有易培养、无毒、可食用等特点，因而其产生的共轭亚油酸具有较高的安全性。

1.3.3 γ-氨基丁酸

γ-氨基丁酸（γ-aminobutyric acid, GABA）是一种在动物、植物和微生物内广泛存在的非蛋白质组成的天然氨基酸。[27] γ-氨基丁酸由谷氨酸经谷氨酸脱羧酶催化而来，在整个神经系统中具有重要作用，可介导哺乳动物中枢神经系统 40% 以上的抑制性神经传导。[28] γ-氨基丁酸具有调节血压、促进食欲、增强免疫力、增强脑活力、调节脂质代谢、增加生长激素分泌、预防肥胖和改善氧化应激等多种功效。[29]

1.3.4 胞外多糖

胞外多糖（Exopolysaccharides, EPS）是乳酸菌分泌到胞外环境的一类结构复杂的多糖类化合物，主要包括荚膜多糖和黏多糖，属于乳酸杆菌的次级代谢产物，其化学结构复杂，但对乳酸杆菌的生长具有重要意义。[30] 胞外多糖具有抗氧化、抗肿瘤、菌群调节、免疫调节等多种生理活性。[31]

1.3.5 乳酸菌素

乳酸菌素（Lactein）是乳酸菌代谢过程中合成并分泌到胞外的一类具有抑菌作用的蛋白或多肽，一般无毒无副作用，易被消化道降解，具有良好的热稳定性，且无不良反应。乳酸菌素可穿过致病菌细胞膜形成孔道和抑制细胞壁合成来达到对致病菌的抑菌、溶菌目的。[32] 依据乳酸菌素相对分子质量、热稳定性、组成氨基酸和作用方式等特点可将乳酸菌素分为羊毛硫乳酸菌素和非羊毛硫乳酸菌素。[33]

1.4 乳酸杆菌的益生功能

1.4.1 乳酸杆菌的黏附、定植与菌群调节

乳酸杆菌对肠道内的病原菌具有拮抗作用，它们能阻碍特定病原菌在肠道内的黏附、定植、繁殖和致病途径。乳酸杆菌需在肠道内保持一定的数量，来防止因肠道的快速收缩蠕动而被排除。乳酸杆菌对黏膜表面的黏附能力可为其提供竞争优势，这一点对于维持消化道菌群平衡非常重要。因此，对肠道黏膜的黏附和定植能力是乳酸杆菌在肠道长期生长、繁殖的必要条件。[25]Jensen 等 [34] 证实 hmpref 0536-10633 基因表达可能在 L. reuteri ATCC PTA 6475（与野生型菌株比较）对 Caco-2 细胞起到重要作用。

乳酸杆菌产生的有机酸、细菌素等抑菌成分可通过抑制致病菌的黏附、定植和繁殖来维持肠道菌群平衡和宿主正常的生理机能。乳酸杆菌可能的菌群调剂机制是：① 产生乳酸、乙酸等短链脂肪酸来降低肠道 pH 值，从而抑制不耐酸性环境的致病菌的生长和繁殖；② 营养竞争，使致病菌缺乏生长所需的某种营养；③ 降低肠腔的氧化还原电位；④ 产生抑菌特殊的抑菌物质，如 H_2O_2 和乳酸菌素；⑤ 增强宿主的免疫活力。[10, 35, 36]Valenzuela 等 [37] 对非洲产狼尾草发酵液中分离获得的植物乳杆菌 2.9 对蜡样芽孢杆菌、大肠杆菌 O157 和沙门氏菌的抑菌试验研究发现，该菌在 8h 对蜡样芽孢杆菌、24h 对沙门氏菌、48h 对大肠杆菌出现显著的抑制效果。

1.4.2 乳酸杆菌与腹泻

近年来，在益生菌的众多生理功能方面的研究中，对腹泻的防治是探讨最深入、应用最广泛的领域。研究证实，采用某些特定的乳酸杆菌来治疗和预防腹泻是非常有用的。乳酸杆菌在急性（轮状病毒）腹泻 [38]、抗生素相关腹泻 [39]、艰难梭菌相关腹泻 [40] 和旅行者腹泻 [41] 表现出较好的效果。益生菌专家曾联合声称，利用益生菌干预急性腹泻已经是被广泛认可的有效方法。[42] 乳酸杆菌缓解腹泻可能的机制是：① 降低肠道 pH，抑制沙门氏

菌等致病菌的生长与繁殖；② 乳酸杆菌及其代谢产物能够抑制大肠杆菌等细菌在消化道上皮细胞的黏附与定植；③ 刺激宿主减少白细胞介素-6（IL-6）、增加免疫球蛋白A（IgA）的产生，来增加宿主免疫活力。[43]

Nagashima等[44]分离日本北海道泡菜的植物乳杆菌Hokkaido对小牛腹泻的影响后发现，该菌株能够降低腹泻在小牛中的发生率，并且在维持肠道菌群平衡方面有较好的表现。但是，Henryk等[45]利用长双歧杆菌、鼠李糖乳杆菌、植物乳杆菌对78位5个月到16岁抗生素相关性腹泻的儿童进行治疗，他们谨慎得出结论，这三株菌对腹泻的影响并不明显。

1.4.3 乳酸杆菌与便秘

便秘是一类由大肠运动异常导致的复杂病症，主要表现为排便困难、便量减少、直肠膨胀或排空时间延长。[46]饮食因素是引起便秘的常见原因之一。恢复肠道菌群平衡和膳食纤维摄入有助于缓解便秘。益生菌对肠道菌群的调节作用是双向的，最终目的是达到肠内微生态菌群平衡。Bekkali等[47]利用包含植物乳杆菌在内的诸多混合益生菌对20位4~16岁儿童的便秘研究发现，该混合菌对儿童的便秘产生了积极的效果。

1.4.4 乳酸杆菌与免疫调节

近几年，乳酸杆菌对免疫系统的调节作用受到了广泛的关注。[48]乳酸杆菌可以通过四条途径刺激机体产生免疫反应：① 作为益生菌在宿主肠道表面黏附、定植并生长繁殖，这些菌体达到一定数目后可以刺激宿主机体免疫系统；② 促进巨噬细胞和树突状细胞对益生菌或裂解产物的吞噬作用，增强巨噬细胞活性；③ 通过调节胃肠道中微生物菌群组成来间接影响免疫系统；④ 可提高肠道内壁黏液细胞表面的局部抗体（如IgA）水平。人和动物的肠道内存在着非常发达的免疫系统，机体免疫系统由多种器官和不同类型的细胞组成。总的来说，益生菌可以在细胞免疫和体液免疫两方面来调节机体免疫水平。[49]孙进等[50]探讨了植物乳杆菌Lp6在小鼠小肠派伊尔结内的细菌源因素，并分析了该益生菌对免疫系统的作用机制。结果显示，该菌可以不同程度地增强小鼠腹腔巨噬细胞的吞噬活性并抑制脾和派伊尔结淋巴细胞增殖。

1.4.5 乳酸杆菌与炎症性肠病

炎症性肠病是一种消化道的炎症,临床表现为腹痛、脓血样大便、肠痉挛等症状。该病主要包括克罗恩病(CD)、溃疡性结肠炎(UC)和结肠袋炎(pouchitis),其发病机制还不完全明确。[51]大肠杆菌炎性肠病被认为是肠道内某个部位发生异常或不受控制的免疫应答所致。与那些没有该病的人群相比较,溃疡性肠炎和克罗恩病患者患结肠癌的风险会增加约 20 倍。在已进行的多项研究中,益生菌在动物体内对这类炎症表现出良好的治疗效果。

1.4.6 乳酸杆菌与口腔健康

口腔作为微生态系统的重要组成部分,其中含有 300 多种细菌。口腔的微生态平衡对于防止口腔疾病与维持口腔健康具有重要意义。现代牙科研究认为,益生菌可用于龋齿、牙周病、口臭、口腔念珠菌病等口腔疾病的治疗。[52]益生菌可黏附于牙齿组织,作为细菌生物膜的一部分,竞争性抑制龋齿细菌和牙周致病菌的生长。[53]

杨颖等[54]利用植物乳杆菌 HO-69 在模拟口腔中研究发现,该菌由于疏水与静电作用介导的较强的黏膜黏附能力对于舌背、颊面及肠道黏膜表面具有较好的黏附性。证实植物乳杆菌 HO-69 满足作为益生菌在口腔中应用的基本条件,可作为潜在的口腔微生态调节剂。

1.4.7 乳酸杆菌与胃损伤修复

幽门螺旋菌(*Helicobacter pylori*,HP)感染是造成胃炎、胃溃疡甚至胃癌的主要发病因素之一。传统的抗生素对幽门螺旋菌有着 90% 的杀灭作用,但其代价较高,并且会使细菌产生抗生素抵抗的副作用。[55, 56]益生菌正在成为一种更加低廉、可大规模应用的制剂用来取代抗生素对胃部疾病进行治疗。

Rokka 等[57]利用 7 种植物乳杆菌对幽门螺旋菌在体外做抑菌试验后发现,这些植物乳杆菌均能表现出对幽门螺旋菌的抑制作用,但菌株间也表现出较大的差异。另外一些研究表明,益生菌及其代谢产物对幽门螺旋菌的侵染及活力具有抑制作用,但还需要更多、设计更严密的试验加以证实。[5]

1.4.8 乳酸杆菌与肠道易激综合征

患肠道易激综合征的人，表现为腹痛、腹胀、便秘或者腹泻等肠道功能失调。肠道易激综合征可能的病因很多，如食物过敏、乳糖不耐症、感染、营养不良、激素代谢失调、小肠细菌过度生长等。[51]目前，对于该类病症，主流医学尚未阐明发病机制且药物不能彻底治愈。从肠道菌群平衡角度来讲，肠道易激综合征患者肠道内乳酸杆菌和双歧杆菌的数量较低，肠道菌群处于失衡状态。目前，益生菌作为最常用的肠道菌群调节手段，在针对肠道易激综合征患者的干预中达到了一定的效果。[58]

1.4.9 乳酸杆菌与代谢综合征

胆固醇对哺乳动物具有重要生理功能，但过高的胆固醇水平会诱导产生高脂血症、动脉粥样硬化、冠心病、甚至糖尿病。以健康人群胆固醇水平为基准，血清胆固醇水平每下降1%，患心血管类疾病的风险便降低2%~3%。[59,60]世界卫生组织（WHO）预计，到2030年大约有2300万人会死于心血管疾病。[61]但是，因常用药物的高价格和副作用，乳酸杆菌成了对付代谢综合征很好的选择。[62]乳酸杆菌在维持肠道菌群平衡，降低胆固醇含量，促进胆汁酸的早期解离，增加胆盐水解酶活力方面扮演着重要的角色[63-65]。

岳喜庆等[66]利用植物乳杆菌1003对小鼠的血清胆固醇作用的研究表明，植物乳杆菌降胆固醇的作用可能是使胆盐由结合态转化为游离态，与胆固醇形成复合物共同沉淀下来，从而使动物血清胆固醇含量下降。

1.4.10 乳酸杆菌与乳糖不耐症

在世界各地的人群中普遍存在乳糖消化不良的问题，这是由于小肠不能完全消化的乳糖在到达结肠后细菌发酵引起的。乳糖不耐症会因为消化道内乳糖分解酶的缺乏而产生各种胃肠不适症状，如腹胀、腹泻或肠胃胀气等症状。全球有超过60%的人有患乳糖不耐症的风险。

对于内源性乳糖酶缺失的人群而言，含乳酸杆菌的酸奶因具有 β- 半乳糖苷酶使得乳糖更容易消化。因此，摄入含乳酸杆菌的酸奶、膳食补充剂，通常可以解决此类问题。[17]

1.4.11 乳酸杆菌的其他功能

益生菌在促进营养物质吸收[67]、保护肝肾健康[68]、改善骨质疏松[69]、延缓衰老[70]、控制内毒素血症[71]、缓解自闭症和减轻焦虑、维护吸烟者健康等方面也有部分效果。但是在这些方面，针对乳酸杆菌的研究较少，需要科研工作者进一步证实。

1.5 乳酸杆菌的安全性

益生菌在作为（功能）食品、膳食补充剂、药品方面具有很长的安全使用历史。[72]益生菌对各类症状的干预效果与多种因素有关，如益生菌的种类、给药剂量和给药方法等。[73]但是，益生菌也存在一定的潜在危害，如可能产生条件性感染、炎性反应和转移抗抗生素基因等状况。但是，在大量有关益生菌应用的研究中，关于乳酸杆菌应用过程中产生的危害则鲜有报道。

报道显示，人体唾液中分离的植物乳杆菌 NCIMB8826 在小鼠体内不能穿越肠道屏障而引起菌体易位。相反，患有结肠炎的小鼠灌胃植物乳杆菌 NCIMB8826 后，可减少内源微生物的易位。[74]SD 大鼠静脉注射植物乳杆菌 299，4 天后这些大鼠死亡，它们的心脏和血液中未检出该菌株的存在，结果证实该菌株是安全的。[75]另外，在菌血症的发病案例中没有发现乳酸杆菌的存在，多种临床研究也证实了乳酸杆菌在人体中使用的安全性。

1.6 前景与挑战

当前，乳酸杆菌在科学研究、营养健康领域已取得了广泛的应用。然而，对乳酸杆菌未来发展的预估依然是困难的，但在现有研究的基础上，我们依然可以预见，在未来的一段时间里，研究者将通过自然选择或基因改造的方式来针对疾病、代谢紊乱、营养与药理需求方面开发特定的乳酸杆菌资源。乳酸杆菌将在：① 作为药物、酶、激素、微量营养物质的载台；② 毒素封存；③ 致癌物质解毒；④ 抗体制备；⑤ 治疗酶不足等方面开发其潜在应用价值，[76] 并需要从分子生物学角度阐明其预防和治疗疾病的机制。

总之，在科研工作者的努力下，随着投入力度的加大，乳酸杆菌一定会为人类做出更大的贡献。

1.7 植物乳杆菌

作为乳酸杆菌中的一种，植物乳杆菌具有环境适应性强、肠道耐受水平高、肠道黏附和定植能力强等特点[14]。此外，该菌在促进肠道菌群平衡、免疫调节、降低胆固醇[62]、缓解乳糖不耐症及抑制肿瘤细胞形成等方面具有重要作用。植物乳杆菌在食品和工业发酵以及医疗保健等领域都有着广泛的应用。

1.8 发酵胡萝卜汁

发酵食品是一类利用由乳酸菌、酵母等有益微生物加工制造的色、香、味俱佳的一类特色食品，如奶酪[77]、酸奶[78]、酒类[79]、酱油[80]、醋[81]、泡菜[82]、开菲尔[83]、发酵果蔬汁[84] 等，该类食品一般具有风味独特、易于吸收等特点。

胡萝卜是二年生伞形科植物，因其富含胡萝卜素、维生素（A、D、E、C、K）和矿物质（钙、钾、铁）等多种营养成分，又被誉为"东方小人参"。[85] 文献显示，β 胡萝卜素具有抗氧化活性，[86, 87] 但其在抗癌和防止动脉粥样硬化方面的结论还存在争议。[88, 89]

胡萝卜汁是一种富含 β 胡萝卜素的常见果蔬汁。[90] 利用益生菌发酵的胡萝卜汁，不仅保存了其自然芳香，而且增加了发酵的风味、营养和口感，比目前市场上单纯的果蔬汁更具市场价值。[91]

1.9 代谢组学

代谢组学（Metabolomics）是近 20 年来兴起并发展起来的一门新的"组学"技术。该组学研究的是生物体或细胞受到内在或外在因素干预条件下产生的内源性小分子标记物（相对分子质量小于 1 000）的代谢变化，并能对代谢相关通路的关键标记物进行定性和定量分析。[92, 93]

在分析技术手段方面，代谢组学主要运用核磁共振（NMR）、质谱（MS）、色谱（HPLC、GC）、色谱质谱联用（LC/MS、GC/MS、UPLC-Q-TOF/MS）技术等定性与定量分析血液、尿液等样本内的小分子代谢物的变化，通过主成分分析、聚类分析等模式分析方法，找出可能的生物标记物（biomarker），并通过通路分析来探索代谢物和机体生理生化变化的关系。[94, 95]

目前，代谢组学已经成为系统生物学的重要组成部分，在疾病诊断、药理毒理分析、营养学、微生物学研究等诸多领域得到广泛应用。[96, 97]

1.10 本书研究的主要内容和创新点

1.10.1 研究价值及意义

目前，针对植物乳杆菌的研究主要集中在单一的功能评价方面，对于它各种益生功能及其发酵产物的综合性评价研究较少。在前期研究中，本实验室从泡菜中筛选得到的一株在高密度培养[98]、胃肠环境耐受[99]、抑制致病菌[100]、果蔬发酵[101]等方面均具有良好性能的植物乳杆菌 NCU116，本研究拟在其调节肠道菌群，缓解便秘，降低血清胆固醇，抑制结肠炎症状和降血糖等方面进行综合评价，探讨其可能的益生特性。

1.10.2 主要研究内容

本书在前期研究的基础上，以 NCU116 为研究对象，主要在以下几个方面进行进一步的研究。

（1）通过不同剂量的植物乳杆菌 NCU116 给小鼠灌胃 5 周来评估该益生菌对肠道生态平衡的作用，并利用气相色谱法检测结肠粪便短链脂肪酸组成，试剂盒检测胆固醇、甘油三酯、氧化应激和免疫相关细胞因子来初步探讨肠道菌群调节对各指标的影响。

（2）利用洛哌丁胺诱导来建立小鼠便秘模型，并灌胃不同剂量植物乳杆菌 NCU116，研究该菌对便秘小鼠粪便重量、含水率、短链脂肪酸含量、小肠推进率、结肠病理与结肠间质细胞 $c\text{-}kit$ 基因的免疫组织化学表达，探讨其对便秘的缓解作用。

（3）利用高脂高胆固醇饲料喂养大鼠诱导脂代谢紊乱动物模型，给予大鼠灌胃不同剂量植物乳杆菌 NCU116，检测其对高脂饮食大鼠抑制高脂血症的影响，并检测该菌对血脂水平（总胆固醇、甘油三酯、高密度脂蛋白胆固醇、低密度脂蛋白胆固醇）、激素水平、肝功能、氧化应激、胰腺与脂肪病理、胆固醇与脂质代谢基因表达等方面的影响。

（4）通过三硝基苯磺酸建立结肠炎小鼠模型，通过灌胃植物乳杆菌 NCU116 及其发酵胡萝卜汁来探讨其对结肠炎小鼠的缓解作用。通过检测胡萝卜汁发酵产物（短链脂肪酸），并对小鼠结肠表观与微观形态、细胞因子、氧化应激和粪便短链脂肪酸等方面对结肠炎损伤修复能力进行评估，得出植物乳杆菌 NCU116 及其发酵胡萝卜汁对结肠炎小鼠可能的作用机制。

（5）利用高脂高糖诱导结合小剂量 STZ 方法建立 II 型糖尿病大鼠模型，通过检测植物乳杆菌 NCU116 及其发酵胡萝卜汁对糖尿病大鼠的血糖、血脂、激素变化的影响。同时，利用病理学方法分析植物乳杆菌 NCU116 及其发酵胡萝卜汁对 II 型糖尿病大鼠胰腺与肾脏的保护作用，并检测粪便短链脂肪酸含量，分析植物乳杆菌 NCU116 及其发酵胡萝卜汁对 II 型糖尿病大鼠糖、脂代谢相关基因的调节作用。

（6）基于 UPLC-Q-TOF/MS 分析技术，研究植物乳杆菌 NCU116（及其发酵胡萝卜汁）对高脂血症和 II 型糖尿病大鼠血液中代谢物小分子的影响，通过 METLIN 数据库进行检索分析，对代谢物小分子进行结构鉴定，利用 Agilent MassHunter 与 Mass Profiler Professional 软件对已鉴定的小分子目标化合物与代谢途径的相关性进行分析，探索植物乳杆菌 NCU116（及其发酵胡萝卜汁）在高脂血症和 II 型糖尿病大鼠血液中的特征生物标志物并进行生物学解释。

1.10.3 技术路线图

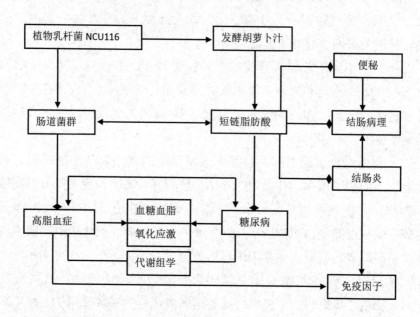

图 1-1　技术路线图（Figure1-1　Technology roadmap）

1.10.4 本书的主要创新点

（1）建立系统的植物乳杆菌 NCU116 对肠道功能、炎症模型与代谢综合征的评价方法。

（2）通过植物乳杆菌 NCU116 对便秘、高脂饮食、结肠炎、糖尿病等方面的影响探讨该益生菌可能的益生特性及其作用机制。

（3）基于 UPLC-QTOF-MS 及 METLIN 等代谢组学手段，分析植物乳杆菌 NCU116 在以高脂血症和糖尿病为代表的代谢综合征大鼠血液中的代谢物信息及代谢紊乱标记物。

第2章 植物乳杆菌NCU116对小鼠肠道菌群的调节作用

2.1 引 言

人的肠道中栖息着复杂而庞大的微生物菌落，其种类可能超过500~1000种，其数量达到了惊人的10^{14}个[102]，基因总数是人体基因的100倍[103]。随着肠道微生态学研究的深入，研究学者逐渐认识到这些细菌是肠道微生态环境必需的组成部分。肠道菌群与人体以共生或拮抗的关系相互依存，它们利用宿主肠道内未消化的食物作为营养，不断地生长繁殖和被排出，来达到肠道菌群的动态平衡维持人体的健康。[104]正常的肠道菌群主要由拟杆菌、双歧杆菌、肠球菌、乳酸杆菌、梭菌、真杆菌等菌属构成，其中95%以上为专性厌氧菌[105]。正常的肠道菌群在提高食物消化率，产生短链脂肪酸、维生素等营养物质，抵御致病菌入侵，调节免疫力等方面具有积极的作用。[106, 107]

肠道菌群可以简单地分为益生菌、致病菌和条件致病菌。当益生菌和致病菌在肠道的生态部位处于动态平衡时，机体会处于健康状态。但是，当微生物菌群失衡的时候，会发生一系列疾病，如腹泻、便秘、肥胖、甚至糖尿病。

益生菌在发酵乳品的安全应用方面已有很长的历史。现今，随着益生菌研究和应用的深入，人们逐渐认识到益生菌的促健康作用。由于人体70%的免疫能力来自肠道，且益生菌在肠道稳态与促进肠道健康方面的巨大作用，因而直接使用益生菌作用于肠道具有重要的研究意义。研究发现，益生菌具有平衡肠道菌群、缓解肠道吸收障碍（如腹泻、乳糖不耐症和肠道易激综合征）的作用。[108]此外，益生菌具有的免疫活性使其具有抑制致病菌生长、防止术后并发症、缓解炎性肠病和防癌的功能。[109-111]

对于所有种类的微生物来说，乳酸菌可能算是与人类关系最密切的一种。[112]本章拟通过对小鼠灌胃不同剂量的植物乳杆菌NCU116来评估该益生菌对肠道生态平衡的作用。

2.2 实验部分

2.2.1 实验材料

2.2.1.1 实验动物

昆明小鼠，SPF级，雄性，20 ± 2 g，40只，购自湖南斯莱克景达实验动物有限公司，许可证号：SCXK（湘）2009-0004。动物饲料由南昌大学医学院实验动物中心提供。

饲养环境：温度23 ± 1℃，湿度55 ± 5%，光暗周期为12 h/12 h 光照黑暗交替进行，实验前适应饲养1周，自由饮食饮水。

本实验动物操作获得南昌大学动物实验伦理委员会许可。

2.2.1.2 实验菌种

植物乳杆菌NCU116菌种（南昌大学食品科学与技术国家重点实验室保藏）。

2.2.1.3 试剂耗材

乳杆菌选择性培养基（LBS琼脂）、BBL琼脂培养基（双歧杆菌）、叠氮钠-结晶紫-七叶苷琼脂（肠球菌）、伊红美兰琼脂EMB（肠杆菌），北京路桥技术有限责任公司；乙酸、丙酸、正丁酸等标准品，上海阿拉丁试剂公司；超氧化物歧化酶（SOD）、谷胱甘肽过氧化物酶（GSH-Px）、过氧化氢酶（CAT）、丙二醛（MDA）试剂盒，南京建成生物工程研究所；胆固醇试剂盒、甘油三酯试剂盒，北京北化康泰临床试剂有限公司；TNF-α、IL-6、IL-10和IL-12酶联免疫试剂盒（ELISA法），武汉博士德生物工程有限公司；其他试剂均为国产分析纯。

2.2.2 实验设备

SHP-150生化培养箱，上海森信实验仪器有限公司；JJ-CJ-ZFD洁净工作台，吴江市净化设备总厂；TGL-16G-A 离心机，上海安亭科学仪器厂；YP10002 马头牌电子天平，上海光正医疗仪器有限公司；AL104 型电子天平，上海梅特勒-托利多仪器公司；Millipore 超纯水机，美国 Millipore 公司；6890N 气相色谱仪、FID 检测器、HP-INNOWAX 色谱柱，美国 Agilent Technologies 公司；Varioskan Flash 全波长多功能酶标仪，美国 Thermo Scientific 公司。

2.2.3 实验方法

昆明小鼠 40 只，20 ± 2 g，雄性。实验环境适应 1 周后，随机分为 4 组，正常组（Normal）、植物乳杆菌 NCU116 低剂量（NCU116-L，1.0×10^7 CFU/mL）组、植物乳杆菌 NCU116 中剂量（NCU116-M，1.0×10^8 CFU/mL）组、植物乳杆菌 NCU116 高剂量（NCU116-H，1.0×10^9 CFU/mL）组。植物乳杆菌 NCU116 悬浮于生理盐水中，正常组灌胃同体积生理盐水，每天每只按 10 mL/kg 剂量灌胃，持续给药 5 周。各组小鼠自由饮食、饮水。

灌胃结束后，无菌收集粪便，眼眶采血后脱椎处死，立即无菌取结肠、肝脏等器官，置于灭菌的离心管中，-80 ℃保存备用。

2.2.3.1 灌胃菌体计数

取植物乳杆菌 NCU116 灌胃悬液，梯度稀释，利用乳杆菌选择性培养基在 37 ℃培养 48 h，计数。

2.2.3.2 体重变化

实验过程中，每周称量并记录体重，并依据体重调整灌胃剂量。

2.2.3.3 植物乳杆菌 NCU116 对肠道菌群的影响

取每只小鼠粪便，按 10 倍系列稀释至合适的浓度。用灭菌的移液器移取稀释液，滴入相应培养皿，分别倾注于各培养基培养（培养条件及方法见表 2-1）。每个样品取 3 个稀释浓度，每个稀释浓度做 3 个平行。经菌落特性与革兰氏染色等特征鉴定并计数菌落。

表2-1 肠道主要菌群的计数方法

Table 2-1 Count method of intestinal flora

粪便菌群	培养基	氧气条件	温度	时间	鉴定方法
乳杆菌	乳杆菌选择性培养基	好氧	37 ℃	48 h	白色或乳白色菌落，G+无芽孢杆菌
双歧杆菌	BBL琼脂培养基	厌氧	37 ℃	48 h	计数白色或乳白色菌落，G+
肠球菌	叠氮钠-结晶紫-七叶苷琼脂	好氧	37 ℃	24 h	计数有明显褐色圈，G+球菌所有菌落
肠杆菌	伊红美兰琼脂EMB	好氧	37 ℃	48 h	典型菌落，G-杆菌

G+：表示革兰氏染色阳性；G-：表示革兰氏染色阴性。

2.2.3.4 植物乳杆菌NCU116对小鼠产短链脂肪酸的影响

分别取结肠粪便、肝脏等样品，制备10%匀浆，离心取上清液，过0.22 μm滤膜。取乙酸、丙酸、正丁酸标准品，配制一定梯度的标准溶液，并用气相色谱仪测相关短链脂肪酸含量。气相色谱条件见表2-2。

表2-2 气相色谱仪工作参数

Table 2-2 GC instrumental parameters

条件	工作参数
色谱柱	HP-INNOWAX（30 m × 0.32 mm I.D）柱
色谱柱温度，进样口温度	240 ℃
检测器	FID，240 ℃
载气及流速	氮气，19.0 mL/min
氢气流速	30 mL/min
空气流速	300 mL/min
进样体积	1 μL
升温程序	100 ℃，0.5min；100-180 ℃，4 ℃/min
时间	20.5 min

2.2.3.5 胆固醇和甘油三酯

采血后离心得血清,严格按照生化试剂盒操作说明进行检测血清中总胆固醇(TC)和甘油三酯(TG)含量。

2.2.3.6 免疫相关细胞因子

血清 TNF-α、IL-6、IL-12、IL-10 等采用酶联免疫法(ELISA),具体操作方法严格按照试剂盒说明书进行。

2.2.3.7 氧化应激

血清中 SOD、GSH-Px、CAT 等酶活力和 MDA 含量的检测严格按照试剂盒说明书进行。

2.2.3.8 统计学分析

各实验组数据以平均数 ± 标准差($\bar{x} \pm s$)表示,采用 SPSS 17.0 软件进行数据统计分析,Duncan's 多重范围检验。$P < 0.05$ 表示组间具有显著性差异,具有统计学意义。

2.3 结果与分析

2.3.1 小鼠体重变化

由图 2-1 可知,在持续给药喂养的过程中,各组小鼠体重都有所增长。灌胃 1 周后,体重在 24.24 ~ 26.05 g 范围内。灌胃 5 周后,植物乳杆菌 NCU116 组体重均比正常组高,其中高剂量组体重最高,达到 34.22 g。提示植物乳杆菌 NCU116 可能具有促生长的作用,但各组无显著性差异($P > 0.05$)。

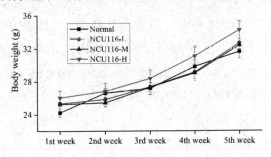

图 2-1 各组体重变化情况

Figure 2-1 Body weight of the four groups

Normal：正常组，灌胃生理盐水；NCU116-L：植物乳杆菌 NCU116 低剂量组（1.0×10^7 CFU/mL）；NCU116-M：植物乳杆菌 NCU116 中剂量组（1.0×10^8 CFU/mL）；NCU116-H：植物乳杆菌 NCU116 高剂量组（1.0×10^9 CFU/mL）。结果以平均数 ± 标准差表示（$n = 10$）。Normal: normal mice treated with saline; NCU116-L: normal mice treated with 10^7 CFU/mL L. plantarum NCU116 group; NCU116-M: normal mice treated with 10^8 CFU/mL L. plantarum NCU116 group; NCU116-H: normal mice treated with 10^9 CFU/mL L. plantarum NCU116 group. Data are expressed as the means ± SEM (n = 10).

2.3.2 小鼠肠道菌群

表2-3 植物乳杆菌NCU116对小鼠肠道菌群的影响（log10 CFU/g 湿重）

Table 2-3 Effect of L. plantarum NCU116 treatment on intestinal microbiota (log10 CFU/g wet contents)

Species	Lactobacillus	Bifidobacterium	Enterobacteriaceae	Enterococcus
Normal	7.55 ± 0.04^a	7.74 ± 0.04^a	5.94 ± 0.08^b	5.67 ± 0.11^c
NCU116-L	9.01 ± 0.15^b	8.64 ± 0.09^b	5.89 ± 0.18^{ab}	5.35 ± 0.04^b
NCU116-M	9.18 ± 0.11^{bc}	9.51 ± 0.04^c	5.79 ± 0.11^{ab}	4.76 ± 0.09^a
NCU116-H	9.41 ± 0.16^c	9.50 ± 0.05^c	5.57 ± 0.05^a	5.16 ± 0.09^b

Normal：正常组，灌胃生理盐水；NCU116-L：植物乳杆菌 NCU116 低剂量组（1.0×10^7 CFU/mL）；NCU116-M：植物乳杆菌 NCU116 中剂量组（1.0×10^8 CFU/mL）；NCU116-H：植物乳杆菌 NCU116 高剂量组（1.0×10^9 CFU/mL）。结果以平均数 ± 标准差表示（$n = 10$），不同字母表示组间具有显著性差异（$P < 0.05$）。Normal: normal mice treated with saline; NCU116-L: normal mice treated with 10^7 CFU/mL L. plantarum NCU116 group; NCU116-M: normal mice treated with 10^8 CFU/mL L. plantarum NCU116 group; NCU116-H: normal mice treated with 10^9 CFU/mL L. plantarum NCU116 group. Data are expressed as the means ± SEM (n = 10).Values within a column with different letters are significantly different ($P < 0.05$).

如表 2-3 所示，实验动物连续灌胃受试物 5 周后，与正常组相比，乳酸杆菌水平在植物乳杆菌 NCU116 各剂量组具有显著性增加（$P < 0.05$），提示经植物乳杆菌 NCU116 干预后，动物肠道内乳酸杆菌水平能够保持在较高水平。相似地，双歧杆菌水平也有较为显著的增加（$P < 0.05$）；结果显示，植物乳杆菌 NCU116 对消化道内双歧杆菌的增殖具有促进作用。在肠杆菌指标方面，与正常组相比，植物乳杆菌 NCU116 的低、中剂量组没有显著性差异（$P > 0.05$），但植物乳杆菌 NCU116 高剂量组能够较好地降低肠杆菌水平。在肠球菌指标方面，与正常组相比（5.67），植物乳杆菌 NCU116 各组能够显著降低肠球菌水平（$P < 0.05$），其中植物乳杆菌 NCU116 中剂量组效果最为明显（4.76）。结果表明：植物乳杆菌 NCU116 连续摄入后，肠道内以乳酸杆菌和双歧杆菌为代表的益生菌能够有效增多，以肠杆菌和肠球菌为代表的致病菌能够有效降低，显示植物乳杆菌 NCU116 能够促进肠道菌群平衡。

2.3.3 小鼠粪便和肝脏短链脂肪酸水平

表2-4 短链脂肪酸标准曲线

Table 2-4 Standard curve of short chain fatty acids (SCFA)

Fatty acids	Equation of standard curves	R2
Acetic acid	[c]=0.010043Y+0.27282	0.9999
Propionic acid	[c]=0.004939Y+0.20393	0.9995
Butyric acid	[c]=0.0035898Y+0.21280	0.9999

[c] 为浓度（μmol/g），Y 是气相图谱的峰面积。

由表 2-4 可知，乙酸、丙酸、正丁酸等短链脂肪酸标品的浓度与峰面积有较好的线性关系，显示该检测方法能够较为准确地测定样品中的短链脂肪酸含量。

表 2-5 显示，在给药 5 周后，与正常组相比，植物乳杆菌 NCU116 各组结肠粪便乙酸、总短链脂肪酸与肝脏乙酸、总短链脂肪酸含量均显著增

加（$P < 0.05$），并具有较好的浓度依赖性；结肠粪便中丙酸和正丁酸含量较正常组也有所增加。但植物乳杆菌NCU116各组肝脏丙酸水平与正常组相比没有显著性差异。结果证实：植物乳杆菌NCU116能够提升结肠粪便和肝脏的短链脂肪酸总体水平。

表2-5 小鼠粪便与肝脏短链脂肪酸含量（μmol/g）

Table 2-5 Short chain fatty acids content in feces and liver (μmol/g)

Samples	Fatty acids	Acetic acid	Propionic acid	Butyric acid	Total SCFA
Feces	Normal	33.54 ± 2.30[a]	3.53 ± 0.29[a]	7.77 ± 0.51[a]	44.83 ± 2.23[a]
	NCU116-L	47.09 ± 3.86[b]	5.48 ± 0.77[b]	9.14 ± 0.46[ab]	61.72 ± 3.34[b]
	NCU116-M	49.52 ± 4.82[b]	5.25 ± 0.66[b]	9.69 ± 0.49[b]	64.46 ± 4.47[b]
	NCU116-H	50.69 ± 2.64[b]	6.35 ± 0.27[b]	9.89 ± 0.45[b]	66.93 ± 2.59[b]
Liver	Normal	8.34 ± 0.49[a]	2.08 ± 0.04	2.19 ± 0.05[ab]	12.61 ± 0.46[a]
	NCU116-L	12.22 ± 0.64[b]	2.02 ± 0.03	2.08 ± 0.07[a]	16.31 ± 0.64[b]
	NCU116-M	14.34 ± 0.67[b]	2.07 ± 0.06	2.23 ± 0.05[ab]	18.64 ± 0.69[b]
	NCU116-H	17.57 ± 1.27[c]	2.08 ± 0.05	2.33 ± 0.07[b]	21.97 ± 1.31[c]

Normal：正常组，灌胃生理盐水；NCU116-L：植物乳杆菌NCU116低剂量组（1.0×10^7 CFU/mL）；NCU116-M：植物乳杆菌NCU116中剂量组（1.0×10^8 CFU/mL）；NCU116-H：植物乳杆菌NCU116高剂量组（1.0×10^9 CFU/mL）。结果以平均数 ± 标准差表示（$n = 10$），不同字母表示组间具有显著性差异（$P < 0.05$）。Normal: normal mice treated with saline; NCU116-L: normal mice treated with 10^7 CFU/mL *L. plantarum* NCU116 group; NCU116-M: normal mice treated with 10^8 CFU/mL *L. plantarum* NCU116 group; NCU116-H: normal mice treated with 10^9 CFU/mL *L. plantarum* NCU116 group. Data are expressed as the means ± SEM ($n = 10$). Values within a column with different letters are significantly different ($P < 0.05$).

2.3.4 小鼠血脂水平

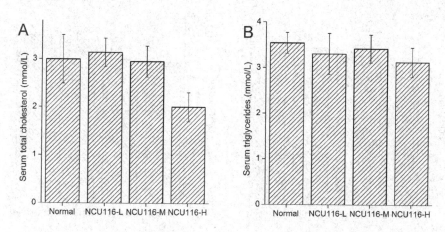

图 2-2 小鼠血清胆固醇（A）和甘油三酯（B）水平（mmol/L）

Figure 2-2 Serum cholesterol (A) and triglyceride (B) levels in mice (mmol/L)

Normal：正常组，灌胃生理盐水；NCU116-L：植物乳杆菌 NCU116 低剂量组（1.0×10^7 CFU/mL）；NCU116-M：植物乳杆菌 NCU116 中剂量组（1.0×10^8 CFU/mL）；NCU116-H：植物乳杆菌 NCU116 高剂量组（1.0×10^9 CFU/mL）。结果以平均数 ± 标准差表示（$n = 10$）。Normal: normal mice treated with saline; NCU116-L: normal mice treated with 10^7 CFU/mL *L. plantarum* NCU116 group; NCU116-M: normal mice treated with 10^8 CFU/mL *L. plantarum* NCU116 group; NCU116-H: normal mice treated with 10^9 CFU/mL *L. plantarum* NCU116 group. Data are expressed as the means ± SEM ($n = 10$).

图 2-2 显示，植物乳杆菌 NCU116 高剂量组能够降低血清胆固醇水平。另外，植物乳杆菌 NCU116 各组能够降低血清甘油三酯水平，但各组之间没有显著性差异（$P > 0.05$）。该结果显示，植物乳杆菌 NCU116 具有一定的降低血清胆固醇和甘油三酯水平的能力。

2.3.5 小鼠氧化应激水平

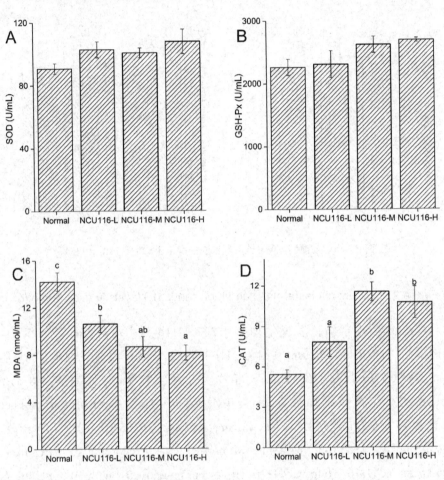

图 2-3 植物乳杆菌 NCU116 对小鼠氧化应激水平的影响

Figure 2-3 Effect of *L. plantarum* NCU116 treatment on oxidative stress in mice

Normal：正常组，灌胃生理盐水；NCU116-L：植物乳杆菌 NCU116 低剂量组（1.0×10^7 CFU/mL）；NCU116-M：植物乳杆菌 NCU116 中剂量组（1.0×10^8 CFU/mL）；NCU116-H：植物乳杆菌 NCU116 高剂量组（1.0×10^9 CFU/mL）。结果以平均数 ± 标准差表示（$n = 10$），不同上标字母表示组间具有显著性差异（$P < 0.05$）。Normal: normal mice treated with saline; NCU116-L: normal mice treated with 10^7 CFU/mL *L. plantarum*

NCU116 group; NCU116-M: normal mice treated with 10^8 CFU/mL *L. plantarum* NCU116 group; NCU116-H: normal mice treated with 10^9 CFU/mL *L. plantarum* NCU116 group. Data are expressed as the means ± SEM (n = 10). Values with different letters are significantly different (P < 0.05).

图2-3显示，植物乳杆菌NCU116能够升高SOD、GSH-Px和CAT活力，降低MDA含量，其中CAT活力中、高剂量组和MDA含量的三个剂量组具有显著性差异（P < 0.05）。结果显示，植物乳杆菌NCU116具有潜在的降低氧化应激损伤的能力。

2.3.6 小鼠血清细胞因子水平

表2-6 小鼠血清细胞因子水平（pg/mL）

Table 2-6 Serum cytokines in mice (pg/mL)

Cytokines	TNF-α	IL-6	IL-10	IL-12
Normal	372.60 ± 18.76[b]	886.68 ± 23.81	482.11 ± 16.68[a]	737.60 ± 63.00
NCU116-L	337.56 ± 5.47[b]	785.35 ± 51.01	579.76 ± 48.08[ab]	664.67 ± 61.66
NCU116-M	290.62 ± 8.52[a]	808.64 ± 58.63	607.63 ± 51.33[b]	686.20 ± 71.72
NCU116-H	298.57 ± 15.33[a]	798.52 ± 26.71	634.96 ± 21.19[b]	623.20 ± 42.63

Normal：正常组，灌胃生理盐水；NCU116-L：植物乳杆菌NCU116低剂量组（1.0×10^7 CFU/mL）；NCU116-M：植物乳杆菌NCU116中剂量组（1.0×10^8 CFU/mL）；NCU116-H：植物乳杆菌NCU116高剂量组（1.0×10^9 CFU/mL）。结果以平均数 ± 标准差表示（n = 10），不同上标字母表示组间具有显著性差异（P < 0.05）。Normal: normal mice treated with saline; NCU116-L: normal mice treated with 10^7 CFU/mL *L. plantarum* NCU116 group; NCU116-M: normal mice treated with 10^8 CFU/mL *L. plantarum* NCU116 group; NCU116-H: normal mice treated with 10^9 CFU/mL *L. plantarum* NCU116 group. Data are expressed as the means ± SEM (n = 10). Values within a column with different letters are significantly different (P < 0.05).

表 2-6 显示，植物乳杆菌 NCU116 能够有效降低促炎因子（TNF-α、IL-6、IL-12）和升高抑炎因子（IL-10）含量，并在高剂量时对 TNF-α 和 IL-10 具有显著性影响（$P < 0.05$）。结果显示，植物乳杆菌 NCU116 可能具有一定的调节免疫能力。

2.4 讨 论

健康的肠道菌群与人和动物机体健康状况具有密切的联系，菌群与宿主机体间的微生态平衡可以确保宿主正常的生理功能和较高的免疫活力，同时肠道菌群还为宿主机体的生长提供大量的营养物质。[113]

在针对动物的生长性能方面，肠道菌群可以产生消化酶类来帮助机体消化淀粉、蛋白质和脂肪等营养成分；[114]益生菌产生的乳酸、乙酸等酸类物质可以降低肠道的酸碱度，酸化的肠道环境有利于矿质元素和维生素 D 的吸收，从而提高饲料的利用率。但是，本实验中，植物乳杆菌 NCU116 表现的促生长能力并不明显。

在肠道微生物菌群与机体之间构成的微生态系统中，优势菌种含量对微生态平衡具有决定性作用。研究表明，益生菌可以建立和维持正常的肠道优势菌群。给失衡的肠道菌群补充微生态制剂（如双歧杆菌、乳酸杆菌），可有助于平衡的恢复。[115]研究认为，菌群在肠上皮表层黏附与肠道定植具有相关性。益生菌可以与肠道内有害微生物黏附和定植位点的竞争，来达到抑制病原菌黏附和定植的目的。此外，益生菌在肠道内还能发酵或代谢产生多种活性化合物（如有机酸和细菌素），从而直接作用于微生物区系，来抑制致病菌。[116]本实验发现，植物乳杆菌 NCU116 表现出促进乳酸杆菌和双歧杆菌等益生菌的生长与繁殖，抑制肠杆菌和肠球菌的生长的性能，提示该菌具有一定的菌群调节能力。

肠道菌群可酵解肠道不易消化的纤维素、半纤维素等物质，能产生短链脂肪酸，并且超过 95% 的有机酸会被宿主吸收和代谢。一般认为，短链脂肪酸与结肠细胞代谢与黏膜损伤修复[117]、生长分化、上皮细胞运输、维持电解质平衡、肝脏脂代谢等生理反应具有重要关系。[21]本研究发现，植物乳杆菌 NCU116 可以显著提高结肠粪便与肝脏短链脂肪酸总体水平，提示植物乳杆菌NCU116可能在结肠黏膜损伤修复方面具有重要意义，我们将在后面章节做进一步探讨。

血脂是指血浆内的中性脂肪和脂类，主要包含甘油三酯、胆固醇和磷脂等。胆固醇和甘油三酯在机体内具有重要的生理作用，但血清中过高的胆固醇水平可能增加血脂紊乱、冠心病、动脉粥样硬化和糖尿病的发病风险。研究表明，乳酸菌具有较好的降低血脂水平的作用，[116]但其机制还没有完全定论。

近年来，促氧化剂（如自由基）与疾病的关系逐渐成为研究热点。尽管益生菌在体内表现出了诸多的生理活性功能，但在抑制氧化应激方面则文献数量有限。[118]然而，现有的文献显示，益生菌的抗氧化功效在体内体外均得到证实，其抑制氧化应激主要从降低体内自由基水平，提高SOD、GSH-Px等抗氧化酶活力，抑制脂质过氧化来减少MDA等方面表现出来。益生菌是一种天然的安全食品，拥有其他抗氧化剂无可比拟的优势。[119]本研究中，植物乳杆菌NCU116表现出潜在的提高抗氧化酶活力与抑制脂质过氧化的作用。

机体免疫系统包括特异性免疫和非特异性免疫。益生菌可以通过肠道菌群、肠道屏障和细胞因子来对机体进行免疫调节，从而使机体降低疾病的发生率。[120]本研究中，植物乳杆菌NCU116使血清中的促炎因子水平降低、抑炎因子水平升高，提示植物乳杆菌NCU116具有一定的免疫调节作用。

2.5　本章小结

本研究通过对不同剂量的植物乳杆菌NCU116对小鼠各项指标研究，表明该菌具有促进乳酸杆菌、双歧杆菌的生长和抑制肠杆菌和肠球菌生长的性能。该菌还表现出在产生短链脂肪酸、降低血脂水平、抑制氧化应激和调节血清细胞因子方面具有一定的作用。但是，其中部分指标不具有显著性差异，该菌的血脂、氧化应激、免疫和菌群调节功能将在后面章节进一步阐述。

第3章 植物乳杆菌NCU116对洛哌丁胺诱导小鼠便秘的缓解作用

3.1 引 言

便秘是一种常见的复杂胃肠疾病，基本症状为排便次数降低、粪便量减少、粪便干结、排出困难、含水量降低等。便秘发病概率为2%~30%，在老年人群的发病概率更高。[121-123] 现已证实，便秘是结肠直肠癌、肠道易激综合征等一些胃肠疾病的风险因子。[124] 便秘的发病因素可能是多方面的，但其机制还不完全清楚。[125] 目前，药物治疗是缓解便秘的首选，但药物的副作用（如心律失常、腹部绞痛、甚至心肌梗死）同样不容忽视。[126]

洛哌丁胺是一种 μ 阿片受体激动剂，常用其诱导痉挛型便秘模型。[127] 该便秘模型具有抑制肠道水分分泌、减少结肠粘膜厚度、降低肠道蠕动、减少水分进入结肠、抑制粪便排空和延迟肠道推进等特点。[124, 128]

近年来，随着对乳酸杆菌在肠道健康领域的深入研究和广泛应用，专家们逐渐认识到乳酸杆菌在缓解和治疗腹泻、炎性肠病和便秘等方面的巨大潜力。[129, 130] 益生菌能够影响结肠蠕动速率、抑制致病菌生长，并且能够促进短链脂肪酸的生成。[131, 132] 短链脂肪酸和乳酸能调节肠道的酸碱度、降低 pH 值，这些因素可能会对结肠蠕动和缓解便秘产生有效的作用。[131]

本研究以植物乳杆菌 NCU116 作为益生菌种[101] 来研究该菌对洛哌丁胺诱导的便秘小鼠症状的缓解作用，并探讨其可能的作用机制。

3.2 实验部分

3.2.1 实验材料

3.2.1.1 实验动物

昆明小鼠，SPF级，雄性，20±2 g，60只，购自湖南斯莱克景达实验动物有限公司，许可证号：SCXK（湘）2009-0004。动物饲料由南昌大学医学院实验动物中心提供。

饲养环境：温度22±1 ℃，湿度55±5%，光暗周期为12 h/12 h光照黑暗交替进行，实验前适应饲养1周，自由饮食饮水。

本实验动物操作获得南昌大学动物实验伦理委员会许可。

3.2.1.2 实验菌种

植物乳杆菌NCU116菌种（南昌大学食品科学与技术国家重点实验室保藏）。

3.2.1.3 试剂耗材

盐酸洛哌丁胺（Loperamide hydrochloride），美国Sigma公司；乙酸、丙酸、正丁酸标准品，上海阿拉丁试剂公司；10%中性福尔马林固定液，南昌市雨露实验器材有限公司；苏木素、伊红染液，碧云天生物技术公司；c-kit（C-19）抗体试剂盒、3% H_2O_2去离子水、山羊血清工作液（试剂A）、生物素化二抗工作液（IgG/Bio，试剂B）、辣根酶标记链霉卵白素工作液（S-A/HRP，试剂C）、DAB显色试剂盒，北京中杉金桥生物技术有限公司；多聚赖氨酸载玻片、Superslip盖玻片，美国Thermo公司；高效切片石蜡（熔点58℃~60℃），上海华灵康复器械厂；无水乙醇、二甲苯、氨水、中性树脂，中国国药集团。

3.2.1.4 主要试剂配制

（1）抗原修复液（0.1 mol/L pH 6.0柠檬酸缓冲液）

A储备液：准确称取29.41 g $C_6H_5Na_3O_7·2H_2O$，加去离子水定容至1000 mL，配制0.1 mol/L柠檬酸三钠溶液；

B储备液：准确称取21 g $C_6H_8O_7·H_2O$，加去离子水定容至1000 mL，

配制 0.1 mol/L 柠檬酸溶液；

准确移取 A 储备液 82 mL 和 B 储备液 18 mL，加入去离子水定容至 1000 mL，配置所需浓度抗原修复液。

（2）PBS 缓冲液

准确称取 8.0 g NaCl，0.2 g KCl，2.084 g $Na_2HPO_4 \cdot 12H_2O$，0.24 g KH_2PO_4，加去离子水定容至 1000 mL，调整 pH 值至 7.4。

3.2.2 实验设备

KD2258 生物组织切片机、KD-T 电脑生物组织摊烤片机、D-BM 生物组织包埋机、KD-BL 包埋机冷冻台、KD-TS3D 生物组织脱水机，浙江科迪仪器设备有限公司；Ti 系列倒置荧光显微镜，日本 Nikon 公司；6890 N 气相色谱仪，美国 Agilent 科技公司；Image Pro Plus 6.0 软件，美国 Media Cybernetics 公司。

3.2.3 实验方法

3.2.3.1 小鼠便秘模型建立

小鼠饲养环境下适应 1 周后，皮下注射 5 mg/kg 盐酸洛哌丁胺（Loperamide hydrochloride）生理盐水悬浊液，每天 9:00 和 18:00 各一次，连续 7 天，其中选取 10 只作为正常组给予同剂量的生理盐水。选取粪便干结、数量减少、重量减轻的小鼠作为便秘小鼠。[133, 134]

3.2.3.2 分组与给药剂量

选取造模成功的 40 只小鼠随机分成模型组（Constipation）、植物乳杆菌 NCU116 低剂量组（NCU116-L, 1.0×10^7 CFU/mL）、植物乳杆菌 NCU116 中剂量组（NCU116-M, 1.0×10^8 CFU/mL）组和植物乳杆菌 NCU116 高剂量组（NCU116-H, 1.0×10^9 CFU/mL），按照 10 mL/kg 的剂量给药。植物乳杆菌 NCU116 悬浮于生理盐水中，平板计数确定菌液浓度。正常组（Normal）和模型组灌胃同剂量生理盐水，连续灌胃 15 天。

3.2.3.3 体重与采食量

实验过程中，称量体重，并测定采食量、饮水量。

3.2.3.4 粪便湿重、干重、含水量[15]

第 15 天给药后，6 h 内观察粪便性状，收集粪便，计数并称重。

收集的粪便置于真空冷冻干燥机内冻干,计算粪便含水量。

含水量(%)=[粪便湿重(g)-粪便干重(g)]÷粪便湿重(g)×100

3.2.3.5 结肠粪便短链脂肪酸含量

参照2.2.3.4步骤进行。

3.2.3.6 小肠推进实验

最后一次给药24 h后,灌胃墨汁,30 min后脱锥处死,打开腹腔,分离小肠肠管,将小肠在无张力情况下拉成直线进行测量肠管长度与墨汁推进长度。墨汁推进率计算公式:

墨汁推进率(%)= 墨汁推进长度(cm)÷ 小肠总长度(cm)×100

3.2.3.7 结肠病理学染色

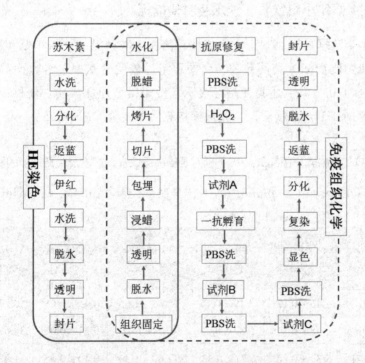

图3-1 苏木素-伊红染色与免疫组织化学流程图

Figure 3-1 The procedures of hematoxylin and eosin staining and immunohistochemistry

取结肠组织并置于10%福尔马林固定液中固定,采用苏木素伊红(hematoxylin and eosin,H&E)染色进行病理学观察,具体流程如图3-1所示。

3.2.3.8 结肠间质细胞c-kit基因的免疫组织化学检测

将结肠组织石蜡包埋样品进行切片并按上述步骤脱蜡水化后，进行c-kit 免疫组织化学染色。具体流程参考图 3-1。

3.2.3.9 统计学分析

各实验组数据以平均数 ± 标准差（$\bar{x} \pm s$）表示，采用 SPSS 17.0 软件进行数据统计分析，Duncan's 多重范围检验。$P < 0.05$ 表示组间具有显著性差异，具有统计学意义。

3.3 结果与分析

3.3.1 便秘小鼠饮食、饮水及粪便指标

表 3-1 数据显示，和正常组相比，洛哌丁胺诱导的便秘小鼠在饮食量、饮水量和粪便指标（粪便粒数、粪便重量、粪便含水率）等方面都有显著降低（$P < 0.05$）。上述粪便指标显示该便秘模型建立成功。便秘小鼠经植物乳杆菌 NCU116 干预后，相关指标均有明显改善（$P < 0.05$）。

表3-1 植物乳杆菌NCU116干预对便秘小鼠饮食、饮水及粪便指标的影响

Table 3-1 Effect of *L. plantarum* NCU116 on food intake, water intake and fecal parameters in constipation mice

Groups	Food intake (g/d)	Water intake (mL/d)	Fecal pellet number	Fecal pellet weight (g)	Fecal moisture (%)
Normal	5.46 ± 0.13[b]	7.19 ± 0.15[c]	15.44 ± 1.06[b]	0.41 ± 0.06[b]	50.99 ± 1.50[b]
Constipation	4.58 ± 0.21[a]	5.87 ± 0.24[a]	8.78 ± 1.42[a]	0.21 ± 0.03[a]	34.65 ± 2.81[a]
NCU116-L	5.33 ± 0.28[b]	6.71 ± 0.33[bc]	12.60 ± 1.39[b]	0.40 ± 0.05[b]	51.87 ± 1.37[b]
NCU116-M	5.29 ± 0.11[b]	6.72 ± 0.19[bc]	14.55 ± 1.40[b]	0.38 ± 0.04[b]	50.04 ± 2.58[b]
NCU116-H	5.25 ± 0.22[b]	7.22 ± 0.19[c]	15.22 ± 1.52[b]	0.41 ± 0.05[b]	49.86 ± 2.49[b]

Normal：正常组；Constipation：便秘模型组；NCU116-L：植物乳杆菌NCU116低剂量组（1.0×10^7 CFU/mL）；NCU116-M：植物乳杆菌NCU116中剂量组（1.0×10^8 CFU/mL）；NCU116-H：植物乳杆菌NCU116高剂量组（1.0×10^9 CFU/mL）。结果以平均数 ± 标准差表示（$n = 10$），不同上标字母表示组间具有显著性差异（$P < 0.05$）。Normal: non-constipation group; Constipation: un-treated constipation group; NCU116-L: constipation treated with 10^7 CFU/mL $L.\ plantarum$ NCU116 group; NCU116-M: constipation treated with 10^8 CFU/mL $L.\ plantarum$ NCU116 group; NCU116-H: constipation treated with 10^9 CFU/mL $L.\ plantarum$ NCU116 group. Results are expressed as the means ± SEM ($n = 10$). Values within a column with different letters are significantly different ($P < 0.05$).

3.3.2 便秘小鼠肠道推进率

表3-2数据显示，与正常组相比，便秘模型组墨汁在30 min 内推进长度与肠道推进率显著降低（$P < 0.05$）；便秘模型组肠道长度有所降低，但结果不具有显著性差异。灌胃植物乳杆菌NCU116后，墨汁推进长度与肠道推进率显著提高（$P < 0.05$）。另外，植物乳杆菌NCU116小肠长度较便秘模型组有所增长，但不具有显著性差异。

表3-2 植物乳杆菌NCU116对便秘小鼠肠道推进率的影响

Table 3-2 Effect of $L.\ plantarum$ NCU116 on intestinal transit ratio in constipation mice.

Groups	Intestinal transit length (cm)	Intestinal length (cm)	Intestinal transit ratio (%)
Normal	36.30 ± 2.27[b]	51.50 ± 0.89	70.57 ± 4.33[b]
Constipation	24.60 ± 1.59[a]	48.70 ± 0.78	50.40 ± 2.89[a]
NCU116-L	36.18 ± 1.64[b]	50.73 ± 1.10	71.51 ± 3.33[b]
NCU116-M	37.27 ± 3.49[b]	51.55 ± 1.59	72.37 ± 6.22[b]
NCU116-H	37.67 ± 2.07[b]	50.89 ± 1.06	73.83 ± 3.31[b]

Normal：正常组；Constipation：便秘模型组；NCU116-L：植物乳杆菌NCU116低剂量组（1.0×10^7 CFU/mL）；NCU116-M：植物乳杆菌NCU116中剂量组（1.0×10^8 CFU/mL）；NCU116-H：植物乳杆菌NCU116高剂量组（1.0×10^9 CFU/mL）。结果以平均数 ± 标准差表示（$n = 10$），同列不同上标字母表示组间具有显著性差异（$P < 0.05$）。Normal: non-constipation group; Constipation: un-treated constipation group; NCU116--L: constipation treated with 10^7 CFU/mL L. plantarum NCU116 group; NCU116-M: constipation treated with 10^8 CFU/mL L. plantarum NCU116 group; NCU116-H: constipation treated with 10^9 CFU/mL L. plantarum NCU116 group. Results are expressed as the means ± SEM ($n = 10$).Values within a column with different letters are significantly different ($P < 0.05$).

3.3.3 便秘小鼠粪便短链脂肪酸

由图3-2和图3-3数据可知，便秘模型组乙酸、丙酸、正丁酸含量较正常组都有所降低，pH值显著升高，其中乙酸、正丁酸、总短链脂肪酸含量及pH值具有显著性差异（$P < 0.05$）。在三个植物乳杆菌NCU116组中，乙酸、丙酸和总短链脂肪酸含量均显著提高，pH值显著降低（$P < 0.05$），显示植物乳杆菌NCU116剂量组能够有效提高粪便短链脂肪酸的含量。并且，乙酸和总短链脂肪酸含量在三个剂量组中呈现一定的剂量效应关系。

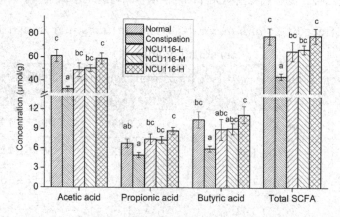

图3-2　各组小鼠粪便乙酸、丙酸、丁酸和总短链脂肪酸的含量

Figure 3-2 Concentration of acetic acid, propionic acid, butyric acid, total SCFA in the mice feces of different groups.

Normal：正常组；Constipation：便秘模型组；NCU116-L：植物乳杆菌 NCU116 低剂量组（1.0×10^7 CFU/mL）；NCU116-M：植物乳杆菌 NCU116 中剂量组（1.0×10^8 CFU/mL）；NCU116-H：植物乳杆菌 NCU116 高剂量组（1.0×10^9 CFU/mL）。总短链脂肪酸 = 乙酸 + 丙酸 + 丁酸。结果以平均数 ± 标准差表示（n = 10），不同上标字母表示组间具有显著性差异（$P < 0.05$）。Normal: non-constipation group; Constipation: un-treated constipation group; NCU116-L: constipation treated with 10^7 CFU/mL *L. plantarum* NCU116 group; NCU116-M: constipation treated with 10^8 CFU/mL *L. plantarum* NCU116 group; NCU116-H: constipation treated with 10^9 CFU/mL *L. plantarum* NCU116 group. Total SCFA = Acetic acid + Propionic acid + Butyric acid. Data are expressed as the means ± SEM (n = 10). Values with different letters are significantly different ($P < 0.05$).

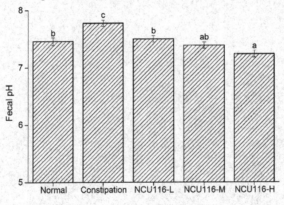

图 3-3　各组小鼠粪便 pH 水平

Figure 3-3 pH values in the mice feces of different groups.

Normal：正常组；Constipation：便秘模型组；NCU116-L：植物乳杆菌 NCU116 低剂量组（1.0×10^7 CFU/mL）；NCU116-M：植物乳杆菌 NCU116 中剂量组（1.0×10^8 CFU/mL）；NCU116-H：植物乳杆菌 NCU116 高剂量组（1.0×10^9 CFU/mL）。结果以平均数 ± 标准差表示（n = 10），不同上标字母表示组间具有显著性差异（$P < 0.05$）。Normal: non-constipation group; Constipation: un-treated constipation group; NCU116-L: constipation treated with 10^7 CFU/mL *L. plantarum* NCU116 group; NCU116-M: constipation treated with 10^8 CFU/mL *L. plantarum* NCU116 group; NCU116-H: constipation treated with 10^9 CFU/mL *L. plantarum* NCU116 group. Results are expressed as the means ± SEM ($n = 10$). Values with different letters are significantly different ($P < 0.05$).

3.3.4 便秘小鼠结肠组织病理学

图 3-4 显示的是植物乳杆菌 NCU116 对远端结肠病理损伤的修复作用。该 H&E 染色显示正常组结肠粘膜完整，排列整齐，杯状细胞丰富，无炎性细胞浸润。便秘模型组黏膜糜烂，杯状细胞减少，炎性细胞浸润。给予植物乳杆菌 NCU116 干预 15 天后，便秘小鼠肠道黏膜结构趋于完整，杯状细胞增多，炎性细胞浸润症状减轻。结果显示，植物乳杆菌 NCU116 能够有效抑制黏膜损伤，减轻炎性细胞浸润。

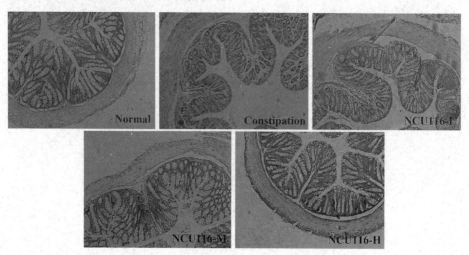

图 3-4 结肠组织 H&E 染色病理学观察

Figure 3-4 Histological findings in the colon with hematoxylin and eosin staining

Normal：正常组；Constipation：便秘模型组；NCU116-L：植物乳杆菌 NCU116 低剂量组（1.0×10^7 CFU/mL）；NCU116-M：植物乳杆菌 NCU116 中剂量组（1.0×10^8 CFU/mL）；NCU116-H：植物乳杆菌 NCU116 高剂量组（1.0×10^9 CFU/mL）。$n = 10$，100 X。Normal: non-constipation group; Constipation: un-treated constipation group; NCU116-L: constipation treated with 10^7 CFU/mL *L. plantarum* NCU116 group; NCU116-M: constipation treated with 10^8 CFU/mL *L. plantarum* NCU116 group; NCU116-H:

constipation treated with 10^9 CFU/mL *L. plantarum* NCU116 group.

3.3.5 便秘小鼠肠道 c-kit 基因表达

正常的 *c-kit* 阳性免疫组化表达为棕色。便秘模型组较正常组结肠 *c-kit* 分布明显减少。由图 3-5 可知，植物乳杆菌 NCU116 灌胃后，*c-kit* 阳性细胞较便秘模型组有所增加。光密度分析表达结果显示，*c-kit* 在便秘模型组中表达较正常组显著降低（$P < 0.05$），灌胃植物乳杆菌 NCU116 后，各剂量组 *c-kit* 表达水平均显著提高（$P < 0.05$）。

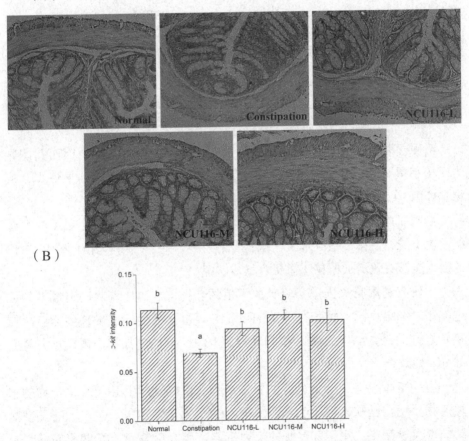

图 3-5 结肠间质细胞（ICC）的免疫组织化学表达（A, 200×）与光密度值（B）

Figure 3-5 Immunohistochemical staining of the colonic cells of Cajal (ICC) by *c-kit* (A, 200×) and the *c-kit* intensity in different groups (B).

Normal：正常组；Constipation：便秘模型组；NCU116-L：植物乳杆菌 NCU116 低剂量组（1.0×10^7 CFU/mL）；NCU116-M：植物乳杆菌 NCU116 中剂量组（1.0×10^8 CFU/mL）；NCU116-H：植物乳杆菌 NCU116 高剂量组（1.0×10^9 CFU/mL）。结果以平均数 ± 标准差表示（$n = 10$），同行不同上标字母表示组间具有显著性差异（$P < 0.05$）。Normal: non-constipation group; Constipation: un-treated constipation group; NCU116-L: constipation treated with 10^7 CFU/mL *L. plantarum* NCU116 group; NCU116-M: constipation treated with 10^8 CFU/mL *L. plantarum* NCU116 group; NCU116-H: constipation treated with 10^9 CFU/mL *L. plantarum* NCU116 group. Data are expressed as the means ± SEM ($n = 10$). Values with different letters are significantly different ($P < 0.05$).

3.4 讨 论

洛哌丁胺具有抑制肠道蠕动、增加粪便排空时间等特性，因而其诱导的便秘小鼠具有较少粪便粒数、重量等特点。[128, 133] 本实验结果证实，植物乳杆菌 NCU116 具有通便和促进肠道蠕动的作用。

洛哌丁胺能够减少饮食饮水量的机制还不清楚，但便秘小鼠（与正常组比较）的粪便排出情况远大于摄食量的差异。[133, 135] 植物乳杆菌 NCU116 的摄入能够有效增加便秘小鼠摄食饮水状况。

本研究利用肠道推进实验来探索植物乳杆菌 NCU116 对肠道蠕动的促进作用。结果显示，植物乳杆菌 NCU116 各组小肠墨汁推进长度能够比便秘模型组至少快 47 %，提示植物乳杆菌 NCU116 能够加快小肠蠕动并具有部分通便效果。

短链脂肪酸是一类肠道（主要是结肠）菌群发酵复杂碳水化合物的末端产物。现在已知乙酸（C2）、丙酸（C3）、丁酸（C4）在人和哺乳动物结肠内大量存在，并具有重要的生理机能。[21] 研究显示，短链脂肪酸具有一定的通便与修复结肠黏膜损伤的作用。[117]Jeon 等 [132] 利用奶酪发酵的燕麦粉末来研究对洛哌丁胺诱导便秘大鼠大肠短链脂肪酸的影响。结果显示，发酵燕麦摄入的大鼠较便秘模型组大鼠短链脂肪酸提高1.4倍。本实验中，便秘小鼠灌胃植物乳杆菌NCU116后，乙酸、丙酸、正丁酸和总短链脂肪酸水平均有明显提高。研究显示，乙酸含量增加有利于肠道结肠蠕动和排便。[136] 短链脂肪酸总体含量增加能够降低肠道环境的pH值。现已证实，结肠低水平的pH值具有诸多有益特性。[137] 本研究中，便秘小鼠短链脂肪酸含量的升高和pH值降低与植物乳杆菌NCU116的摄入密不可分。

洛哌丁胺诱导的便秘小鼠能够降低结肠黏液层和黏膜厚度，进而导致结肠黏膜损伤。[124] 本研究发现，植物乳杆菌NCU116具有修复结肠黏膜损伤的特性。研究报道，奶酪发酵的燕麦粉末对便秘大鼠结肠黏膜损伤修复可能源于短链脂肪酸的作用。[132]Araki等[138] 报道，对结肠炎大鼠利用膳食纤维和丁酸梭菌联合干预，发现大鼠结肠黏膜隐窝损伤和炎性因子显著受到抑制，可能的作用机制是丁酸梭菌发酵膳食纤维产生的短链脂肪酸对结肠损伤的保护作用。报道显示，短链脂肪酸是结肠上皮细胞保持完整性的必备条件，它们能刺激结肠上皮细胞增值和分化，并可作为肠细胞的主要能量来源。另外，远端结肠较高的丁酸水平能够抑制大肠癌变的发生。[139, 140] 因此，便秘小鼠结肠损伤的修复可能与植物乳杆菌NCU116促进结肠短链脂肪酸水平的显著提高有关。

肠道间质细胞（Interstitial cells of Cajal, ICC）广泛分布在结肠的各肌层中，它在控制肠道蠕动方面发挥着重要的作用。[141, 142] 文献显示，ICC是胃肠道平滑肌的起搏细胞，可以产生生理慢波。该慢波通过促进电活动的扩散来控制消化道中平滑肌有节律的蠕动与收缩。ICC的减少可能导致慢波活动紊乱、减少平滑肌的收缩，进而导致慢传输便秘的发生。另外，便秘个体内ICC的减少可能会破坏正常的结肠蠕动节律。[125, 142] 肠道内ICC的基因表达可以利用$c\text{-}kit$的阳性表达来衡量。[143]Lee等[144] 发现$c\text{-}kit$对ICC网络具有重要作用，$c\text{-}kit$的缺失能够破坏ICC网络，可能导致便秘与巨结肠病的发生。本研究利用$c\text{-}kit$的免疫组织化学方法检测ICC的表达，发现ICC在便秘小鼠结肠中表达降低，植物乳杆菌NCU116摄入后，ICC表达显著提高，其机理暂时还不清楚。

3.5 本章小结

（1）植物乳杆菌 NCU116 对洛哌丁胺诱导的便秘小鼠的粪便指标具有显著的缓解作用。

（2）植物乳杆菌 NCU116 可以明显提高便秘小鼠的肠道蠕动水平。

（3）植物乳杆菌 NCU116 能够促进结肠短链脂肪酸的产生，并促进结肠上皮细胞与黏膜损伤的修复。

（4）植物乳杆菌 NCU116 可改善胃肠道间质细胞（ICC）的 $c\text{-}kit$ 基因表达，从而增加结肠的蠕动性能，促进排便。

第4章 植物乳杆菌NCU116对高脂饮食大鼠抑制血脂紊乱的机制

4.1 引 言

　　血脂代谢紊乱是一类心血管病的风险因子,该类疾病已成为多个国家疾病人群的头号死因。高脂血症是以血液中胆固醇含量偏高为主要特征。[145]报道显示,高水平的血清胆固醇水平可增加冠心病、动脉粥样硬化和糖尿病等疾病的患病风险。[59, 60]但是,当前常规的抑制血脂紊乱的药物价格较高,并且副作用较多。因此,研究有效的药物替代策略来预防心血管病是一个紧迫的任务。[62]

　　近年来,研究发现益生菌在维持肠道菌群平衡与血脂平衡方面具有重要功能。[146, 147]乳酸菌能够通过同化吸收作用降低血清胆固醇水平,[63]并能分泌胆盐水解酶在胆盐的早期解离方面具有重要作用。[64, 65]报道显示,乳酸杆菌在降低血清胆固醇水平方面具有突出表现,但其机制还不完全清楚。

　　本实验拟利用高脂高胆固醇饲料饲养大鼠形成的高脂血症动物模型,同时给大鼠灌胃不同剂量植物乳杆菌NCU116,检测其对高脂饮食大鼠抑制高脂血症的影响;并判定该菌对体重、血脂水平、氧化应激、胆固醇代谢基因表达等方面产生的影响。

4.2 实验部分

4.2.1 实验材料

4.2.1.1 实验动物

Sprague-Dawley(SD)大鼠,SPF级,雄性,120～150 g,40只,购

自北京维通利华实验动物有限公司,许可证号:SCXK(京)2012-0001。动物饲料由南昌大学医学院实验动物中心提供。

饲养环境:温度 23±1 ℃,湿度 55±5%,光暗周期为 12 h/12 h 光照黑暗交替进行,实验前适应饲养 1 周,自由饮食饮水。

本实验动物操作获得南昌大学动物实验伦理委员会许可。

4.2.1.2 实验菌种

植物乳杆菌 NCU116 菌种(南昌大学食品科学与技术国家重点实验室保藏)。

4.2.1.3 试剂耗材

胆固醇(TC)试剂盒、甘油三酯(TG)试剂盒、低密度脂蛋白胆固醇(LDL-C)试剂盒、高密度脂蛋白胆固醇(HDL-C)试剂盒,北京北化康泰临床试剂有限公司;SOD 测定试剂盒、GSH-Px 测定试剂盒、MDA 测定试剂盒、CAT 测定试剂盒、T-AOC 测定试剂盒,南京建成生物工程研究所;苏木精染液、伊红染液,北京中杉金桥生物技术有限公司;cDNA 反转录试剂盒,立陶宛 Thermo 公司;SYBR® Premix Ex Taq™ 试剂盒,日本 Takara 公司;PCR 引物,中国 Invitrogen 公司;胰岛素测定试剂盒、瘦素测定试剂盒、脂联素测定试剂盒,北京华英生物技术研究所。

4.2.2 实验设备

ACCU-CHEK Performa 血糖仪,德国 Roche Diagnostics 公司;CX31 电子显微镜,日本 Olympus 公司;KD-TS3D 生物组织脱水机、KD2258 生物组织切片机、KD-T 电脑生物组织摊烤片机、D-BM 生物组织包埋机,浙江科迪仪器设备有限公司;7900 HT RT-qPCR 系统,加拿大 Applied Biosystems 公司;R-911 全自动放免计数仪,中国科技大学实业总公司。

4.2.3 实验方法

4.2.3.1 实验设计

大鼠实验环境下适应 1 周后,随机被分为正常组(Normal diet, ND)、高脂模型组(High fat diet, HFD)、植物乳杆菌 NCU116 低剂量组(NCU116-L,

10^8 CFU/mL）和植物乳杆菌 NCU116 高剂量组（NCU116-H，10^9 CFU/mL），每组 10 只，给所有大鼠正常饮水，给予高脂模型组、植物乳杆菌 NCU116 低剂量组和植物乳杆菌 NCU116 高剂量组高脂饲料；给予正常组正常饲料。[148] 按照 10 mL/kg 体重给高剂量组和低剂量组灌胃相应剂量的植物乳杆菌 NCU116，高脂模型组和正常组灌胃同等剂量生理盐水，连续 5 周。

4.2.3.2 体重、饮食、饮水测定

实验过程中，定期称量体重，记录饮食、饮水情况。

4.2.3.3 脂代谢水平测定

每周眼球取血，离心得血清。测定血清 TC、TG、HDL-C 和 LDL-C 水平。

4.2.3.4 胰岛素、瘦素、脂联素水平测定

灌胃结束后，禁食过夜，心脏刺穿采血，离心得血清。放免法测定胰岛素（Insulin）、瘦素（Leptin）、脂联素（Adiponectin）含量。

4.2.3.5 口服葡萄糖耐量试验

动物禁食不禁水 12 h，尾静脉取血（0 min）测定血糖，给动物灌胃 2 g/kg BW 剂量葡萄糖，分别在 30 min、60 min、90 min、120 min 测定血糖值。[149]

4.2.3.6 氧化应激水平

参照 2.2.3.7 操作。

4.2.3.7 胰腺与脂肪组织病理学观察

实验结束后，大鼠经麻醉，分离胰腺、脂肪等组织，称重。组织 H&E 染色，显微镜下观察病理学变化，并拍照记录。具体操作按 3.2.3.7 进行。

4.2.3.8 RT-qPCR 测定脂代谢基因表达

根据 TRIzol 试剂说明书提取总 RNA，反转录得 cDNA。β-actin 作为参比基因，利用 7900 HT Fast Real-Time PCR System 测定 low density lipoprotein (LDL) receptor, 3-hydroxy-3-methylglutaryl coenzyme A (HMG-CoA) reductase, cholesterol 7α-hydroxylase (CYP7A1) 等基因的相对定量表达，$2^{-\Delta\Delta C_T}$ 分析数据。PCR 引物序列如表 4-1。

表4-1 实时定量PCR引物列表

Table 4-1 Primers used for quantitative real-time PCR

Gene	Sense primer (5'→3')	Antisense primer (5'→3')
LDL receptor[150]	CAGCTCTGTGTGAACCTGGA	TTCTTCAGGTTCGGGATCAG
HMG-CoA reductase[151]	CCCAGCCTACAAACTGGAAA	CCATTGGCACCTGGTACTCT
CYP7A1[152]	CACCATTCCTGCAACCTTTT	GTACCGGCAGGTCATTCAGT
SREBP-2[151]	AGACTTGGTCATGGGGACAG	GGGGAGACATCAGAAGGACA
β-actin[153]	TGTTGTCCCTGTATGCCTCT	TAATGTCACGCACGATTTCC

4.2.3.9 统计学分析

各实验组数据以平均数 ± 标准差（$\bar{x} \pm s$）表示，采用 SPSS 17.0 软件进行数据统计分析，Duncan's 多重范围检验。$P < 0.05$ 表示组间具有显著性差异，具有统计学意义。

4.3 结果与分析

4.3.1 体重、饮食、饮水变化

表 4-2 显示，实验过程中，4 组动物在饮食饮水方面无明显差异。实验 5 周后，与正常组相比，高脂模型组体重显著升高（$P < 0.05$）；与高脂模型组相比，植物乳杆菌 NCU116 低剂量组（387.11 ± 5.6 g）与高剂量组（380.22 ± 6.76 g）动物体重显著降低（$P < 0.05$）。

表4-2 实验过程中各组动物体重、摄食和饮水变化情况

Table 4-2 Body weight, food intake and water consumption in the four groups

Parameters	Groups	1th week	2nd week	3rd week	4th week	5th week
Body weight (g)	ND	227.58 ± 2.91[a]	263.77 ± 3.88[a]	292.62 ± 5.68[a]	317.43 ± 6.86[a]	336.38 ± 6.92[a]

续表

Parameters	Groups	1th week	2nd week	3rd week	4th week	5th week
Body weight (g)	HFD	246.89 ± 2.64[b]	300.61 ± 2.14[c]	340.93 ± 3.19[c]	378.50 ± 4.41[c]	407.57 ± 4.90[c]
	NCU116-L	244.72 ± 2.82[b]	289.70 ± 3.24[bc]	327.52 ± 4.20[bc]	361.55 ± 5.02[b]	387.11 ± 5.6[b]
	NCU116-H	241.83 ± 4.05[b]	285.29 ± 5.29[b]	321.17 ± 6.03[b]	354.74 ± 6.27[b]	380.22 ± 6.76[b]
Food intake (g/d)	ND	25.58 ± 0.74	26.83 ± 1.27	25.75 ± 1.12	25.87 ± 0.51[b]	25.80 ± 0.79[b]
	HFD	26.51 ± 0.66	27.20 ± 1.48	25.60 ± 1.01	24.39 ± 0.56[ab]	25.49 ± 0.49[ab]
	NCU116-L	25.79 ± 0.48	24.31 ± 0.80	23.61 ± 1.00	23.77 ± 0.59[a]	23.72 ± 0.81[ab]
	NCU116-H	24.70 ± 0.56	24.93 ± 1.16	24.28 ± 0.90	23.75 ± 0.56[a]	23.48 ± 0.59[a]
Water consumption (mL/d)	ND	34.23 ± 0.65	33.87 ± 2.33	35.37 ± 0.97	35.53 ± 1.33	34.83 ± 1.28
	HFD	34.50 ± 1.28	34.17 ± 3.80	35.08 ± 2.02	34.42 ± 2.22	38.38 ± 2.68
	NCU116-L	33.17 ± 0.65	31.57 ± 1.85	31.17 ± 0.75	31.07 ± 0.92	33.67 ± 0.99
	NCU116-H	33.83 ± 1.56	30.73 ± 2.20	31.23 ± 1.27	31.57 ± 1.50	33.23 ± 1.19

ND：正常组；HFD：高脂模型组；NCU116-L：植物乳杆菌 NCU116 低剂量组；NCU116-H：植物乳杆菌 NCU116 高剂量组。结果以平均数 ± 标准差表示（$n = 10$），同列不同上标字母表示组间具有显著性差异（$P < 0.05$）。ND: rats on the normal diet; HFD: rats on the high fat diet; NCU116-L: rats on the high fat diet +10^8 CFU/mL L. plantarum NCU116; NCU116-H: rats on the high fat diet +10^9 CFU/mL L. plantarum NCU116. Data are expressed as the means ± SEM ($n = 10$). Values within a column with different letters are significantly different ($P < 0.05$).

4.3.2 脂代谢水平

实验过程中，每周测定大鼠血清中 TC、TG、HDL-C 和 LDL-C 的水平。正常组各项指标基本保持稳定（表4-3）。第 5 周数据显示，模型组血清 TC（2.51 mmol/L）是正常组（1.15 mmol/L）的两倍多，TG 和 LDL-C 水平较正常组显著升高，HDL-C 显著降低。给予大鼠植物乳杆菌 NCU116 连续灌胃 5 周后，两个剂量组（尤其是高剂量组）的 TC、TG、LDL-C 明显降低（$P < 0.05$），HDL-C 含量有所升高，但结果不具有显著性差异（P

>0.05）。结果显示，植物乳杆菌NCU116具有一定调节血脂的能力。

表4-3 植物乳杆菌NCU116对大鼠脂代谢水平的影响（mmol/L）

Table 4-3 Effect of *L. plantarum* NCU116 treatment on serum lipids levels (mmol/L)

Parameters	Groups	1th week	2nd week	3rd week	4th week	5th week
TC	ND	1.07 ± 0.13^a	1.17 ± 0.08	1.14 ± 0.16^a	1.07 ± 0.11^a	1.15 ± 0.10^a
	HFD	1.40 ± 0.12^b	1.65 ± 0.16	2.13 ± 0.25^b	2.30 ± 0.10^c	2.51 ± 0.08^c
	NCU116-L	1.34 ± 0.10^{ab}	1.61 ± 0.16	1.66 ± 0.14^{ab}	1.92 ± 0.05^{bc}	2.16 ± 0.24^{bc}
	NCU116-H	1.26 ± 0.04^{ab}	1.45 ± 0.29	1.56 ± 0.16^a	1.78 ± 0.21^b	1.84 ± 0.15^b
TG	ND	0.32 ± 0.03^a	0.41 ± 0.03^a	0.34 ± 0.07	0.48 ± 0.04^a	0.49 ± 0.05^a
	HFD	0.53 ± 0.07^b	0.62 ± 0.09^b	0.58 ± 0.10	0.77 ± 0.07^b	0.81 ± 0.06^b
	NCU116-L	0.36 ± 0.03^a	0.48 ± 0.06^{ab}	0.45 ± 0.07	0.63 ± 0.07^{ab}	0.62 ± 0.03^a
	NCU116-H	0.39 ± 0.03^a	0.43 ± 0.06^{ab}	0.45 ± 0.06	0.55 ± 0.07^a	0.58 ± 0.07^a
HDL-C	ND	0.73 ± 0.04	0.76 ± 0.02^b	0.75 ± 0.03^c	0.74 ± 0.02^b	0.75 ± 0.05^b
	HFD	0.64 ± 0.03	0.56 ± 0.03^a	0.54 ± 0.05^a	0.54 ± 0.04^a	0.52 ± 0.04^a
	NCU116-L	0.73 ± 0.03	0.63 ± 0.03^a	0.63 ± 0.02^{ab}	0.60 ± 0.03^a	0.59 ± 0.05^{ab}
	NCU116-H	0.68 ± 0.04	0.64 ± 0.02^a	0.68 ± 0.04^{bc}	0.64 ± 0.07^{ab}	0.68 ± 0.06^{ab}
LDL-C	ND	0.23 ± 0.02^a	0.26 ± 0.06^a	0.25 ± 0.03^a	0.25 ± 0.04^a	0.23 ± 0.04^a
	HFD	0.45 ± 0.04^b	0.73 ± 0.04^b	1.33 ± 0.14^c	1.41 ± 0.09^c	1.52 ± 0.06^c
	NCU116-L	0.36 ± 0.03^b	0.67 ± 0.03^b	0.80 ± 0.03^b	0.93 ± 0.04^b	1.01 ± 0.10^b
	NCU116-H	0.40 ± 0.03^b	0.61 ± 0.06^b	0.64 ± 0.05^b	0.85 ± 0.03^b	0.90 ± 0.05^b

ND：正常组；HFD：高脂模型组；NCU116-L：植物乳杆菌NCU116低剂量组；NCU116-H：植物乳杆菌NCU116高剂量组。结果以平均数±标准差表示（$n = 10$），同列不同上标字母表示组间具有显著性差异（$P < 0.05$）。ND: rats on the normal diet; HFD: rats on the high fat diet; NCU116-L: rats on the high fat diet +10^8 CFU/mL *L. plantarum* NCU116;

NCU116-H: rats on the high fat diet +10⁹ CFU/mL *L. plantarum* NCU116. Data are expressed as the means ± SEM (n = 10). Values within a column with different letters are significantly different (P < 0.05).

4.3.3 口服糖耐量试验

实验结束前，禁食 12 小时，口服 2 g/kg 葡萄糖，分别在 0 min、30 min、60 min、90 min、120 min 时测定血糖。30 min 后，各组血糖均达到峰值，正常组为 7.64 mmol/L，高脂模型组显著提高到 9.80 mmol/L（P < 0.05）。植物乳杆菌 NCU116 低剂量组为 8.46 mmol/L，植物乳杆菌 NCU116 高剂量组为 8.30 mmol/L，两个植物乳杆菌 NCU116 组与模型组相比，均有所降低，但结果无显著差异。在 60 min 到 120 min 过程中，各组血糖均有所下降。

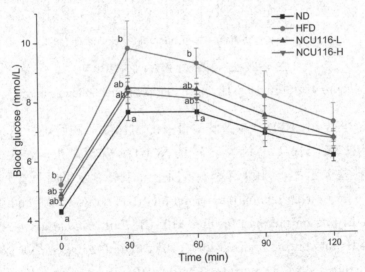

图 4-1　口服葡萄糖后血糖浓度的变化（mmol/L）

Figure 4-1 Blood glucose concentrations after the oral glucose load

ND：正常组；HFD：高脂模型组；NCU116-L：植物乳杆菌 NCU116 低剂量组；NCU116-H：植物乳杆菌 NCU116 高剂量组。结果以平均数 ± 标准差表示（n = 10），图中不同上标字母表示组间具有显著性差异（P < 0.05）。ND: rats on the normal diet; HFD: rats on the high fat diet; NCU116-L: rats on the high fat diet +10⁸ CFU/mL *L. plantarum* NCU116; NCU116-H: rats on the high fat diet +10⁹ CFU/mL *L. plantarum* NCU116.

Data are expressed as the means ± SEM ($n = 10$). Values with different letters are significantly different ($P < 0.05$).

4.3.4 血清激素水平

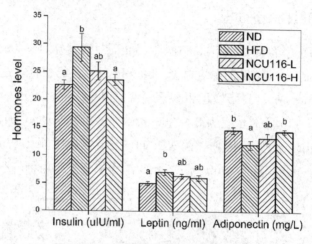

图 4-2 血清胰岛素、瘦素、脂联素水平

Figure 4-2 Serum insulin, leptin, and adiponectin levels in rats

ND：正常组；HFD：高脂模型组；NCU116-L：植物乳杆菌 NCU116 低剂量组；NCU116-H：植物乳杆菌 NCU116 高剂量组。结果以平均数 ± 标准差表示（$n = 10$），图中不同上标字母表示组间具有显著性差异（$P < 0.05$）。ND: rats on the normal diet; HFD: rats on the high fat diet; NCU116-L: rats on the high fat diet +10^8 CFU/mL L. plantarum NCU116; NCU116-H: rats on the high fat diet +10^9 CFU/mL L. plantarum NCU116. Data are expressed as the means ± SEM ($n = 10$). Values with different letters are significantly different ($P < 0.05$).

由图 4-2 可知，高脂模型组血清胰岛素和瘦素含量较正常组显著提高（$P < 0.05$），并且植物乳杆菌 NCU116 高剂量组胰岛素水平较为接近正常组。高脂模型组血清脂联素水平较正常组显著降低（$P < 0.05$），植物乳杆菌 NCU116 灌胃 5 周后，两个剂量组脂联素水平均有所升高，但仅有植物乳杆菌 NCU116 高剂量组与高脂模型组有显著性差异（$P < 0.05$）。

4.3.4 血清氧化应激水平

取血清利用试剂盒测定氧化应激相关指标,与正常组比较,模型组 SOD、GSH-Px、CAT 和 T-AOC 分别显著降低为 133.70 ± 8.71 U/mL、3428.65 ± 115.64 U/mL、9.14 ± 0.72 U/mL 和 3.08 ± 0.56 U/mL ($P < 0.05$),MDA 显著升高到 7.35 ± 0.46 nmol/mL。灌胃植物乳杆菌 NCU116 后,相关指标均能有效改善,其中高剂量组上述指标分别达到 163.82 ± 3.72 U/mL、4031.01 ± 52.28 U/mL、14.51 ± 1.46 U/mL、5.24 ± 0.31 U/mL 和 5.61 ± 0.45 nmol/mL(表 4-4)。实验结果表明,植物乳杆菌 NCU116 能够抑制血清的氧化应激损伤。

表4-4 植物乳杆菌NCU116对大鼠氧化应激水平的影响

Table 4-4 Effect of *L. plantarum* NCU116 treatment on oxidative stress in rats

Oxidative stress	ND	HFD	NCU116-L	NCU116-H
SOD (U/mL)	191.61 ± 6.83^c	133.70 ± 8.71^a	168.66 ± 4.33^b	163.82 ± 3.72^b
GSH-Px (U/mL)	4177.58 ± 101.36^b	3428.65 ± 115.64^a	4002.70 ± 23.61^b	4031.01 ± 52.28^b
MDA (nmol/mL)	4.78 ± 0.41^a	7.35 ± 0.46^c	6.63 ± 0.49^{bc}	5.61 ± 0.45^{ab}
CAT (U/mL)	17.51 ± 1.60^c	9.14 ± 0.72^a	12.93 ± 1.63^{ab}	14.51 ± 1.46^{bc}
T-AOC (U/mL)	7.10 ± 0.76^c	3.08 ± 0.56^a	5.45 ± 0.46^b	5.24 ± 0.31^b

ND:正常组;HFD:高脂模型组;NCU116-L:植物乳杆菌 NCU116 低剂量组;NCU116-H:植物乳杆菌 NCU116 高剂量组。结果以平均数 ± 标准差表示($n = 10$),同行不同上标字母表示组间具有显著性差异($P < 0.05$)。ND: rats on the normal diet; HFD: rats on the high fat diet; NCU116-L: rats on the high fat diet $+10^8$ CFU/mL *L. plantarum* NCU116; NCU116-H: rats on the high fat diet $+10^9$ CFU/mL *L. plantarum* NCU116. Data are expressed as the means ± SEM ($n = 10$). Values within a row with different letters are significantly different ($P < 0.05$).

4.3.5 胰腺和脂肪组织病理学

胰腺和脂肪组织经福尔马林固定、脱水、包埋、切片后,利用伊红和苏木素染色。显微镜下观察,可见正常组胰岛细胞结构清晰完整、大小适中,腺泡细胞规则有序地围绕在胰岛周围。高脂模型组胰岛 β 细胞尺寸变小,胰岛部分空泡样病变,部分细胞萎缩凋亡。灌胃植物乳杆菌 NCU116 后,胰岛损伤减轻,腺泡细胞排列趋于规则(图 4-3A)。

与正常组比较,高脂模型组脂肪细胞明显肥大,同等视野面积内细胞明显减少,并有少量炎性细胞浸润。但是,给予植物乳杆菌 NCU116 后,脂肪细胞肥大症状明显减轻,炎性细胞浸润减少(图 4-3B)。

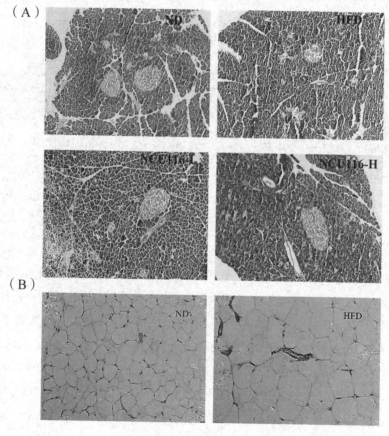

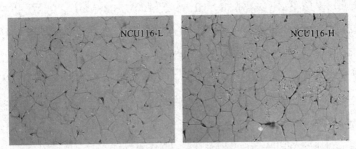

图 4-3 植物乳杆菌 NCU116 对高脂饮食大鼠胰腺（A）和脂肪（B）病理水平的影响（100X）

Figure 4-3 Effects of *L. plantarum* NCU116 on histology of pancreas (A) and adipose tissues (B) in rats fed a high fat diet (100X).

ND：正常组；HFD：高脂模型组；NCU116-L：植物乳杆菌 NCU116 低剂量组；NCU116-H：植物乳杆菌 NCU116 高剂量组。ND: rats on the normal diet; HFD: rats on the high fat diet; NCU116-L: rats on the high fat diet $+10^8$ CFU/mL *L. plantarum* NCU116; NCU116-H: rats on the high fat diet $+10^9$ CFU/mL *L. plantarum* NCU116.

4.3.6 脂代谢基因表达

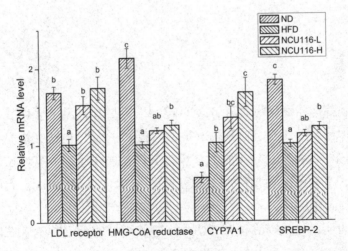

图 4-4 肝细胞 LDL receptor, HMG-CoA reductase, CYP7A1 and SREBP-2 的 mRNA 表达水平

Figure 4-4 Hepatic mRNA expression levels for the LDL receptor, HMG-CoA reductase, CYP7A1 and SREBP-2 genes.

以高脂模型组作为对照，正常组 LDL receptor（低密度脂蛋白受体）、HMG-CoA reductase（3-羟基-3甲基戊二酸单酰辅酶 A 还原酶）、CYP7A1（胆固醇 7α-羟化酶）、SREBP-2（胆固醇调节元件结合蛋白 2）基因分别是模型组的 1.68、2.13、0.57、1.83 倍。灌胃植物乳杆菌 NCU116 后，上述指标明显改善，其中 LDL receptor 在低、高剂量组表达分别是模型组的 1.52 和 1.74 倍，CYP7A1 在低、高剂量组表达分别是模型组的 1.34 和 1.67 倍。实验结果显示，植物乳杆菌 NCU116 通过调控 LDL receptor 和 CYP7A1 基因来缓解高胆固醇症状。

ND：正常组；HFD：高脂模型组；NCU116-L：植物乳杆菌 NCU116 低剂量组；NCU116-H：植物乳杆菌 NCU116 高剂量组。结果以平均数 ± 标准差表示（$n = 10$），图中不同上标字母表示组间具有显著性差异（$P < 0.05$）。ND: rats on the normal diet; HFD: rats on the high fat diet; NCU116-L: rats on the high fat diet +10^8 CFU/mL *L. plantarum* NCU116; NCU116-H: rats on the high fat diet +10^9 CFU/mL *L. plantarum* NCU116. The graph represents the mRNA levels relative to *β*-actin. Data are expressed as the means ± SEM ($n = 10$). Values with different letters are significantly different ($P < 0.05$).

4.4 讨 论

高脂血症通常表现为血清 TC、TG 和 LDL-C 升高，同时 HDL-C 降低，这些指标的异常变化与心血管病有很强的相关性。[154]本研究中，大鼠通过高脂高胆固醇饮食来诱导高脂血症模型的产生，并通过体重增长、血脂代谢、mRNA 表达等方面来探讨植物乳杆菌 NCU116 对高脂饮食大鼠高脂血症症状的抑制作用。

结果表明，经植物乳杆菌 NCU116 连续灌胃 5 周可以明显降低高脂饮食大鼠血清 TC、TG 和 LDL-C 并能提高 HDL-C 水平。报道显示，高脂饮食能够升高血清 TC、TG 和 LDL-C 水平，并具有导致动脉粥样硬化的风险，[155]但高水平的 HDL-C 能够降低该类心血管病的风险。[156]另外，LDL-C 是血脂中的高度危险因子，因为被氧化的低密度脂蛋白胆固醇能够侵入动脉血管壁上并形成动脉粥样硬化斑块。[154,155,157]因此，降低 TC 和 LDL-C 是降低动脉粥样硬化的有效手段。[156,158]

大鼠经过连续的高脂饮食能够在葡萄糖耐受试验中增加血糖水平。高脂喂养诱导的高血糖症状能够诱导胰岛素分泌，增加体重、内脏与皮下脂肪重量，[159]植物乳杆菌 NCU116 能够有效缓解这些症状。此外，激素（如胰岛素、瘦素和脂联素）在参与脂代谢过程中扮演重要角色。有研究表明，胰岛素能够影响瘦素的产生，[160]肥胖的个体具有较高的血清瘦素水平。[161]脂联素由脂肪细胞分泌，在血糖、血脂代谢中具有重要作用。[162]根据本研究结果，瘦素水平的降低和脂联素水平的提高可能与植物乳杆菌 NCU116 摄入能够提高胰岛素敏感度有着重要关系。

研究表明，富含脂肪的饮食可以升高体内自由基的含量，进而诱导产生氧化应激。[163]活性氧含量的增加可引起脂质过氧化，从而导致组织细胞膜、蛋白质和 DNA 受损。[164]尽管导致机体衰老是一个复杂的过程，但氧化应激作为其中的一个关键因素影响着衰老的进程。[165]每个机体都有抵制氧化损伤的体系，它包括 SOD、GSH-Px、CAT 等酶类物质。[166,167]MDA 是脂质过氧化的标志物，它具有 DNA 和蛋白质毒性。[168]本研究发现，摄入植物乳杆菌 NCU116 能够显著提升 SOD、GSH-Px 和 CAT 酶类活力，升高 T-AOC 水平，并降低 MDA 含量。因此，植物乳杆菌 NCU116 可能具有潜在的抵抗氧化损伤的能力。

大鼠连续摄入高脂高胆固醇饲料会导致内脏脂肪堆积，并具有产生胰岛素抵抗、脂代谢紊乱和患心血管病的风险。[169]经植物乳杆菌 NCU116 干预后，病理学观察可知胰腺损伤有所降低，脂肪细胞肥大症状有所缓解，提示植物乳杆菌 NCU116 可能具有抑制脂肪肝和肥胖的功能。

胆固醇体内的相关代谢通路主要有三种：吸收，合成和分泌[170]。本实验中，通过检测肝脏 LDL 受体、HMG-CoA 还原酶、CYP7A1 和 SREBP-2 来探讨植物乳杆菌 NCU116 对胆固醇代谢的影响。LDL 受体是一类广泛分布于肝细胞表面，并能通过转录调节机制来控制低密度脂蛋白胆固醇水平。[171]在高脂模型组中，LDL 受体的基因表达水平显著降低，使血清中低密度脂蛋白胆固醇水平升高。胆汁酸和胆固醇具有抑制 HMG-CoA 还原酶的表达性能。[172]本研究证实，植物乳杆菌 NCU116 能够改善肝细胞 HMG-CoA 还原酶的基因表达。SREBP-2 是一种能调控胆固醇摄入和平衡的基因，它的表达在高脂饮食中受到抑制。[173]植物乳杆菌NCU116可以降低血清胆固醇水平，可能与SREBP-2的表达水平提高有关。另外，高脂喂养诱导CYP7A1表达水平提高。研究证实，CYP7A1的高水平表达能够降低高脂喂养仓鼠的血清胆固醇和甘油三酯水平。[174]CYP7A1是一种微粒体细胞色素p450调节胆汁酸合成代谢的关键限速酶，能调节胆汁酸合成的整体速率。[175]本研究中，对高脂饮食大鼠灌胃植物乳杆菌NCU116，表现出调节CYP7A1基因表达的能力。综上所述，植物乳杆菌NCU116表现的调节血脂的能力可能与上述基因表达水平有关。

4.5　本章小结

（1）植物乳杆菌NCU116对高脂饮食诱导高脂血症大鼠模型具有潜在的调节血脂功效。

（2）植物乳杆菌NCU116对高脂血症大鼠模型的氧化应激损伤有较好的缓解作用。

（3）植物乳杆菌NCU116能够改善高脂饮食大鼠胰腺和脂肪的微观结构，促进机体损伤修复。

（4）植物乳杆菌NCU116具有一定的降血糖能力，但需要进一步探讨。

（5）植物乳杆菌NCU116能够调控脂代谢与胆固醇代谢基因表达，对以高脂血症为代表的代谢综合征的预防有着重要意义。

第5章 植物乳杆菌NCU116对大鼠脂肪肝病变的抑制作用

5.1 引 言

非酒精性脂肪肝病（non-alcoholic fatty liver disease, NAFLD）是指每日饮酒在小于10 g情况下，肝脏脂肪累积大于5%，以肝细胞内脂肪过度沉积和脂肪性病变为主要特征的临床综合征，与肥胖、高脂血症和糖尿病等代谢综合征具有密切关系。[176-178] 从病理学来讲，非酒精性脂肪肝病包含单纯肝脏脂肪变性以及其演变的肝炎和肝硬化。[179, 180] 多篇文献报道，持续的高脂饮食能够影响动物体内肠道菌群平衡和诱导免疫指标发生变化，并发展为肥胖及其并发症，包括非酒精性脂肪肝病[181]。

有研究显示，肠道菌群在肝脏损伤方面有着重要作用。调节肠道菌群能够有效抑制肥胖相关脂肪肝病的发生。[177] 当前，应用益生菌来提升肠道屏障功能，从而干预不同类型的慢性肝损伤已经成为一种可能的选择。[3, 182, 183] 作为益生菌中的一类，乳酸菌在啮齿类动物模型中表现出了较好的缓解非酒精性脂肪肝病的功能，但其中关于肝脏脂质代谢通路还不清楚。[184]

前期研究发现，植物乳杆菌NCU116具有较好的体外性能和体内降低胆固醇的活性。[98, 185] 通过脂代谢相关知识可知，植物乳杆菌NCU116的这些性能可能具有改善非酒精性脂肪肝病的能力。因此，本章节研究拟采用高脂饲料喂养大鼠诱导其产生非酒精性脂肪肝病症状，并探讨植物乳杆菌NCU116对该症状缓解作用及其可能的机制。

5.2 实验部分

5.2.1 实验材料

5.2.1.1 实验动物

Sprague-Dawley（SD）大鼠，SPF级，雄性，120～150 g，40只，购自北京维通利华实验动物有限公司，许可证号：SCXK（京）2012-0001。动物饲料由南昌大学医学院实验动物中心提供。

饲养环境：温度23±1 ℃，湿度55±5 %，光暗周期为12 h/12 h 光照黑暗交替进行，实验前适应饲养1周，自由饮食饮水。

本实验动物操作获得南昌大学动物实验伦理委员会许可。

5.2.1.2 实验菌种

植物乳杆菌NCU116菌种（南昌大学食品科学与技术国家重点实验室保藏）。

5.2.1.3 试剂耗材

SOD测定试剂盒、GSH-Px测定试剂盒、MDA测定试剂盒、CAT测定试剂盒、T-AOC测定试剂盒、谷丙转氨酶（ALT）测定试剂盒、谷草转氨酶（AST）测定试剂盒、总胆红素测定试剂盒，南京建成生物工程研究所；LPS酶联免疫试剂盒、IL-6酶联免疫试剂盒、IL-10酶联免疫试剂盒、TNF-α酶联免疫试剂盒，上海西塘科技有限公司；苏木精染液、伊红染液，北京中杉金桥生物技术有限公司；cDNA反转录试剂盒，立陶宛Thermo公司；SYBR® Premix Ex Taq™，日本Takara公司；PCR引物，中国Invitrogen公司；GLC-463脂肪酸甲酯混标，美国Nu-Chek Prep公司。

5.2.2 实验设备

CX31电子显微镜，日本Olympus公司；KD-TS3D生物组织脱水机、KD2258生物组织切片机、KD-T电脑生物组织摊烤片机、D-BM生物组织包埋机，浙江科迪仪器设备有限公司；7900 HT RT-qPCR系统，加拿大Applied Biosystems公司；Varioskan Flash酶标仪，美国Thermo Scientific

公司；6890N 气相色谱仪、FID 检测器、CP-Sil 88 色谱柱，美国 Agilent Technologies 公司。

5.2.3 实验方法

5.2.3.1 实验设计

大鼠实验环境下适应 1 周后，随机被分为正常组（ND）、高脂模型组（HFD）、植物乳杆菌 NCU116 低剂量组（NCU116-L，10^8 CFU/mL）和植物乳杆菌 NCU116 高剂量组（NCU116-H，10^9 CFU/mL），每组 10 只。所有大鼠正常饮水，给予模型组、高剂量组和低剂量组高脂饲料；给予正常组正常饲料。按照 10 mL/kg 体重给高剂量组和低剂量组灌胃相应剂量的植物乳杆菌 NCU116，高脂模型组和正常组灌胃同等剂量生理盐水，连续 5 周。

实验结束后，大鼠经水合氯醛麻醉，心脏刺穿采血，离心得血清。无菌结肠粪便、肝脏、脂肪、肾脏等备用。肝脏与脂肪器官指数的计算公式为：器官指数 = 器官质量（g）/ 大鼠体重（g）× 100。

5.2.3.2 血清肝功能测定

严格按照试剂盒说明书测定血清谷丙转氨酶（ALT）、谷草转氨酶（AST）活性和总胆红素含量。

5.2.3.3 肝脏氧化应激水平

参照 2.2.3.7 操作。

5.2.3.4 脂多糖与免疫因子测定

利用酶联免疫试剂盒，根据说明书测定血清脂多糖（LPS）、白介素（IL）-6、IL-10 与肿瘤坏死因子（TNF）-α 含量。

5.2.3.5 肝脏病理学染色

取肝脏，按 3.2.3.7 操作。

5.2.3.6 肝脏脂肪酸测定

肝脏放入液氮中充分研磨成粉末。氯仿 - 甲醇萃取法提取粗脂肪，加入 $C_{21:0}$ 内标，经氢氧化钾 - 甲醇溶液甲酯化。[186] 采用气相色谱法测定肝脏脂肪酸组成：CP-Sil 88 型毛细管色谱柱，FID 检测器，进样量为 1 μL，氢气为载气，进样口和检测器温度 250 ℃，恒流模式，分流比为 10:1，氢气流量为 26 mL/min，空气流量为 300 mL/min。升温程序为：初温 60 ℃，保持 5 min，以 11.5 ℃/min 升至 170 ℃，保持 25 min；再以 5 ℃/min 的速率升至 200 ℃，保持 5 min；最后以 2 ℃/min 升温至 215 ℃，保持 20 min。参考相关文献，脂肪酸组成采用内标定量因子计算。[187, 188]

5.2.3.7 RT-qPCR 测定脂代谢基因表达

参照 4.2.3.8 提取粪便与肝脏总 RNA，反转录得 cDNA。利用 7900 HT Fast Real-Time PCR System 测定粪便相关菌群与肝脏脂代谢相关基因（表 5-1）的相对定量表达，$2^{-\Delta\Delta C_T}$ 分析数据。

表5-1 实时定量PCR引物列表

Table 5-1 Primers used for quantitative real-time PCR

Gene	Sense primer (5'→3')	Antisense primer (5'→3')
PPARα	GGTCATACTCGCAGGAAA	AGCAAATTATAGCAGCCAC
PPARγ	TCAGGGCTGCCAGTTTCG	GCTTTTGGCATACTCTGTGATCTC
PPARδ	AACGAGATCAGCGTGCATGTG	TGAGGAAGAGGCTGCTGAAGTT
PGC1α	TGAGAGGGCCAAGCAAAG	ATAAATCACACGGCGCTCTT
CPT1α	CGGTTCAAGAATGGCATCATC	TCACACCCACCACCACGAT
FAS	GTCTGCAGCTACCCACCCGTG	CTTCTCCAGGGTGGGGACCAG
ACC	ACAGAGATGGTGGCTGATGTC	GATCCCCATGGCAATCTG
SCD1	TGCCAGAGGGAATAGGGAAA	CTCTCCCATCCTTACTTACAAACCA
Lactobacillus spp.	AGCAGTAGGGAATCTTCCA	ATTYCACCGCTACACATG
Bifidobacterium spp.	CTCCTGGAAACGGGTGG	GGTGTTCTTCCCGATATCTACA
Enterobacteriaceae spp.	CATTGACGTTACCCGCAGAAGAAGC	CTCTACGAGACTCAAGCTTGC
Bacteroides spp.	ATAGCCTTTCGAAAGRAAGAT	CCAGTATCAACTGCAATTTTA
β-actin	TGTTGTCCCTGTATGCCTCT	TAATGTCACGCACGATTTCC

5.2.3.8 统计学分析

各实验组数据以平均数 ± 标准差（$\bar{x}\pm s$）表示，采用 SPSS 17.0 软件进行数据统计分析，Duncan's 多重范围检验。$P < 0.05$ 表示组间具有显著性差异，具有统计学意义。

5.3 结果与分析

5.3.1 肝功能

如图 5-1 所示，与正常组比较，高脂模型组 ALT 与 AST 活力和总胆红素水平显著升高（$P < 0.05$）。但植物乳杆菌 NCU116 摄入后，三个指标水平均有所降低，其中，植物乳杆菌 NCU116 高剂量组在 AST 活力方面与高脂模型组具有显著性差异。

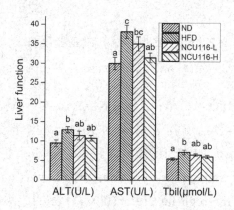

图 5-1　植物乳杆菌 NCU116 对大鼠肝功能的影响

Figure 5-1 Effect of *L. plantarum* NCU116 treatment on liver function in rats

ND：正常组；HFD：高脂模型组；NCU116-L：植物乳杆菌 NCU116 低剂量组，NCU116-H：植物乳杆菌 NCU116 高剂量组。结果以平均数 ± 标准差表示（$n = 10$），图中不同上标字母表示组间具有显著性差异（$P < 0.05$）。ND: rats on the normal diet; HFD: rats on the high fat diet; NCU116-L: rats on the high fat diet $+10^8$ CFU/mL *L. plantarum* NCU116; NCU116-H: rats on the high fat diet $+10^9$ CFU/mL *L. plantarum* NCU116. Data are expressed as the means ± SEM ($n = 10$). Values with different letters are significantly different ($P < 0.05$).

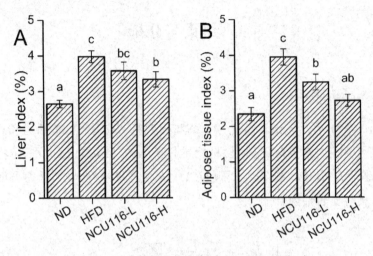

图 5-2　大鼠肝脏指数（A）和脂肪指数（B）

Figure 5-2 Liver (A) and adipose tissue (B) indices of rats

ND：正常组；HFD：高脂模型组；NCU116-L：植物乳杆菌 NCU116 低剂量组；NCU116-H：植物乳杆菌 NCU116 高剂量组。结果以平均数 ± 标准差表示（$n = 10$），图中不同上标字母表示组间具有显著性差异（$P < 0.05$）。ND: rats on the normal diet; HFD: rats on the high fat diet; NCU116-L: rats on the high fat diet $+10^8$ CFU/mL $L.$ $plantarum$ NCU116; NCU116-H: rats on the high fat diet $+10^9$ CFU/mL $L.$ $plantarum$ NCU116. Data are expressed as the means ± SEM ($n = 10$). Values with different letters are significantly different ($P < 0.05$).

5.3.2 肝脏指数与脂肪酸组成分析

图 5-2 显示，高脂饮食诱导的大鼠模型中肝脏和脂肪指数较正常组明显升高（$P < 0.05$）。给予高脂饲料饲养，同时灌胃植物乳杆菌 NCU116，能够有效缓解这一症状。

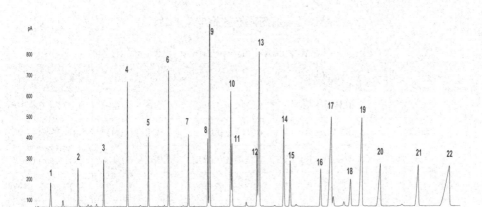

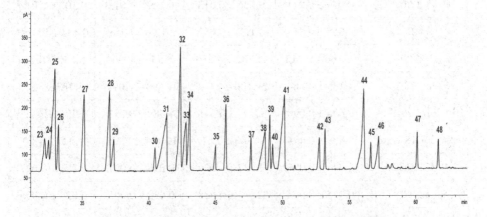

图 5-3 GLC-463 脂肪酸甲酯标准品对应的气相色谱图

Figure 5-3 Chromatogram of fatty acid methyl esters for GLC-463 standard

对应峰序号的名称：1: C4:0，2: C6:0，3: C7:0，4: C8:0，5: C9:0，6: C10:0，7: C11:0，8: C11:1，9: C12:0，10: C12:1，11: C13:0，12: C13:1，13: C14:0，14: C14:1，15: C15:0，16: C15:1，17: C16:0，18: 9t-C16:1，19: 9c-C16:1，20: C17:0，21: C17:1，22: C18:0，23: 9t-C18:1，24: 11t-C18:1，25: 6c+9c-C18:1，26: 11c-C18:1，27: 9t12t-C18:2n-6+C19:0，28: C19:1，29: 9c12c-C18:2n-6，30: 18:3n-6，31: C20:0，32: 5c-C20:1+C18:3n-3，33: 8c-C20:1，34: 11c-C20:1，35: C21:0，36: C20:2n-6，37: C20:3n-6，38: C22:0，39: C20:3n-3，40: C20:4n-6，41: C22:1n-9，42: C22:2n-6，43: C20:5n-3，44: C24:0+C22:3n-3，45: C22:4n-6，46: C24:1n-9，47: C22:5n-3，48: C22:6n-3。

表5-2 肝脏脂肪酸组成（mg/g）

Table 5-2 Fatty acids composition in liver (mg/g)

Fatty acids	ND	HFD	NCU116-L	NCU116-H
C14:0	0.045 ± 0.005^a	0.250 ± 0.062^c	0.165 ± 0.029^b	0.171 ± 0.079^b
9cC14:1	0.006 ± 0.002^a	0.015 ± 0.003^b	0.013 ± 0.001^b	0.017 ± 0.009^b
C15:0	0.029 ± 0.003^a	0.090 ± 0.018^c	0.062 ± 0.014^b	0.059 ± 0.017^b
C16:0	4.585 ± 0.303^a	12.354 ± 2.039^c	9.783 ± 1.302^b	8.972 ± 1.879^b
9cC16:1	0.206 ± 0.045^a	1.287 ± 0.301^c	1.123 ± 0.292^{bc}	0.942 ± 0.328^b
C17:0	0.093 ± 0.011^a	0.136 ± 0.013^b	0.129 ± 0.013^b	0.107 ± 0.012^a
9cC17:1	0.085 ± 0.014^a	0.167 ± 0.051^b	0.133 ± 0.046^{ab}	0.131 ± 0.070^{ab}
C18:0	4.806 ± 0.282	5.333 ± 0.650	5.556 ± 0.946	5.038 ± 0.359
9cC18:1	1.135 ± 0.264^a	24.130 ± 3.945^c	21.793 ± 3.099^{bc}	18.387 ± 3.254^b
11cC18:1	0.606 ± 0.090^a	1.997 ± 0.316^c	1.589 ± 0.251^b	1.496 ± 0.367^b
9c12cC18:2n-6	3.809 ± 0.404^a	14.785 ± 2.615^c	12.109 ± 1.771^b	11.329 ± 1.807^b
6c9c12cC18:3n-6	0.024 ± 0.005^a	0.127 ± 0.029^c	0.098 ± 0.007^b	0.091 ± 0.027^b
C20:0	0.015 ± 0.005^a	0.033 ± 0.008^b	0.032 ± 0.003^b	0.029 ± 0.005^b
9c12c15cC18:3n-3	0.036 ± 0.012^a	0.373 ± 0.057^c	0.302 ± 0.046^b	0.275 ± 0.067^b
11cC20:1	0.024 ± 0.003^a	0.499 ± 0.122^c	0.364 ± 0.095^b	0.308 ± 0.066^b
C20:2n-6	0.062 ± 0.010^a	0.425 ± 0.111^c	0.299 ± 0.067^b	0.285 ± 0.071^b
C20:3n-6	0.187 ± 0.040^a	0.896 ± 0.173^c	0.701 ± 0.124^b	0.656 ± 0.150^b
C22:0	0.035 ± 0.002^a	0.045 ± 0.006^b	0.041 ± 0.009^{ab}	0.042 ± 0.009^{ab}
C20:4n-6	5.651 ± 0.131^c	4.755 ± 0.516^b	4.579 ± 0.599^{ab}	4.186 ± 0.257^a
C20:5n-3	0.084 ± 0.020^a	0.192 ± 0.042^c	0.157 ± 0.024^{bc}	0.154 ± 0.030^b
C24:0	0.093 ± 0.016^a	0.298 ± 0.085^c	0.220 ± 0.067^b	0.198 ± 0.042^b
C22:5n-3	0.221 ± 0.044^a	0.457 ± 0.113^b	0.315 ± 0.075^a	0.298 ± 0.075^a
C22:6n-3	1.723 ± 0.131^a	2.358 ± 0.439^b	1.932 ± 0.385^a	1.759 ± 0.182^a
Σ SFA	9.701 ± 0.531^a	18.539 ± 2.313^c	15.988 ± 1.803^b	14.617 ± 2.037^b

续 表

Fatty acids	ND	HFD	NCU116–L	NCU116–H
∑ MUFA	2.071 ± 0.333^a	28.156 ± 4.500^c	25.083 ± 3.500^{bc}	21.337 ± 4.016^b
∑ PUFA	11.804 ± 0.597^a	24.405 ± 3.663^c	20.520 ± 2.780^b	19.066 ± 2.340^b
∑ FA	23.593 ± 1.390^a	71.197 ± 10.161^c	61.686 ± 7.456^b	55.108 ± 8.078^b

ND：正常组；HFD：高脂模型组；NCU116-L：植物乳杆菌 NCU116 低剂量组；NCU116-H：植物乳杆菌 NCU116 高剂量组。nd，表示未检出；∑SFA，总饱和脂肪酸；∑MUFA，总单不饱和脂肪酸；∑PUFA，总多不饱和脂肪酸；∑FA，总脂肪酸。结果以平均数 ± 标准差表示（$n = 10$），同行不同上标字母表示组间具有显著性差异（$P < 0.05$）。ND: rats on the normal diet; HFD: rats on the high fat diet; NCU116–L: rats on the high fat diet +10^8 CFU/mL L. plantarum NCU116; NCU116–H: rats on the high fat diet +10^9 CFU/mL L. plantarum NCU116. "nd" means not detected. ∑SFA, total saturated fatty acids; ∑MUFA, total monounsaturated fatty acids; ∑PUFA, total polyunsaturated fatty acids; ∑FA: total fatty acids. Data are expressed as the means ± SEM ($n = 10$). Values within a row with different letters are significantly different ($P < 0.05$).

根据图 5-3 脂肪酸标准品图谱分析，得出各组肝脏脂肪酸水平。结果显示，高脂模型组的肝脏脂肪酸水平显著高于正常组（表 5-2）。其中，高脂模型组的 ∑SFA（18.539 mg/g），∑MUFA（28.156 mg/g），∑PUFA（24.405 mg/g）和 ∑FA（71.197 mg/g）分别是正常组的 1.91，13.60，2.07 和 3.02 倍。∑FA 在植物乳杆菌 NCU116 低剂量组和高剂量组分别降低到 61.686 mg/g 和 55.108 mg/g，表明植物乳杆菌 NCU116 在降低肝脏总脂肪酸水平方面有较好表现。

5.3.3 脂多糖与细胞免疫因子水平

图 5-4 表明，高脂模型组大鼠血清 LPS、IL-6 和 TNF-α 水平显著高于正常组，IL-10 水平显著低于正常组（$P < 0.05$）。但经植物乳杆菌 NCU116 干预后，其各项免疫指标均有所缓解。其中，植物乳杆菌 NCU116

高剂量组的各项指标都较为接近正常组,与高脂模型组有显著性差异($P < 0.05$)。

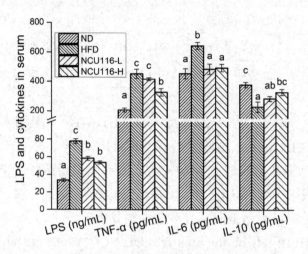

图 5-4　血清脂多糖与细胞免疫因子水平

Figure 5-4 LPS and cytokines in serum

ND:正常组;HFD:高脂模型组;NCU116-L:植物乳杆菌 NCU116 低剂量组;NCU116-H:植物乳杆菌 NCU116 高剂量组。结果以平均数 ± 标准差表示($n = 10$),图中不同上标字母表示组间具有显著性差异($P < 0.05$)。ND: rats on the normal diet; HFD: rats on the high fat diet; NCU116-L: rats on the high fat diet $+10^8$ CFU/mL $L.\ plantarum$ NCU116; NCU116-H: rats on the high fat diet $+10^9$ CFU/mL $L.\ plantarum$ NCU116. Data are expressed as the means ± SEM ($n = 10$). Values with different letters are significantly different ($P < 0.05$).

5.3.4 结肠粪便菌群

如图 5-5 所示,不同给药组的结肠菌群,乳杆菌属、双歧杆菌属、肠杆菌属和拟杆菌属表达水平各有不同。与正常组比较,高脂模型组中乳杆菌属和双歧杆菌属含量下调,肠杆菌属含量上调($P < 0.05$)。但植物乳杆菌 NCU116 剂量组可以上调乳杆菌属、双歧杆菌属基因表达水平,降低拟杆菌属表达水平($P < 0.05$)。

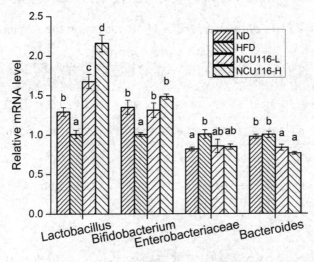

图 5-5 结肠粪便菌群 mRNA 表达水平

Figure 5-5 mRNA expression of colonic bacterial flora

ND：正常组；HFD：高脂模型组；NCU116-L：植物乳杆菌 NCU116 低剂量组；NCU116-H：植物乳杆菌 NCU116 高剂量组。结果以平均数 ± 标准差表示（$n = 10$），图中不同上标字母表示组间具有显著性差异（$P < 0.05$）。ND: rats on the normal diet; HFD: rats on the high fat diet; NCU116-L: rats on the high fat diet +10^8 CFU/mL L. plantarum NCU116; NCU116-H: rats on the high fat diet +10^9 CFU/mL L. plantarum NCU116. Data are expressed as the means ± SEM ($n = 10$). Values with different letters are significantly different ($P < 0.05$).

5.3.5 肝脏氧化应激水平

由表 5-3 可知，与正常组相比，高脂模型组 SOD（18.48 U/mgprot）、GSH-Px（1196.10 U/mgprot）、CAT（42.43 U/mgprot）、T-AOC（0.70 U/mgprot）水平显著降低，MDA（0.89 nmol/mgprot）水平显著升高。给予两个剂量组植物乳杆菌 NCU116 灌胃后，各项指标均有所改善。其中，植物乳杆菌 NCU116 高剂量组各指标均与高脂模型组具有显著性差异（$P < 0.05$）。

表5-3 植物乳杆菌NCU116对大鼠肝脏氧化应激水平的影响

Table 5-3 Effect of *L. plantarum* NCU116 treatment on the response of rats to oxidative stress in liver

Oxidative stress	ND	HFD	NCU116-L	NCU116-H
SOD (U/mgprot)	29.31 ± 1.38^c	18.48 ± 1.99^a	22.85 ± 1.20^{ab}	25.12 ± 2.18^{bc}
GSH-Px (U/mgprot)	1638.63 ± 46.71^b	1196.10 ± 57.91^a	1435.27 ± 98.93^a	1501.60 ± 103.91^a
MDA (nmol/mgprot)	0.31 ± 0.05^a	0.89 ± 0.11^b	0.47 ± 0.07^a	0.45 ± 0.07^a
CAT (U/mgprot)	55.23 ± 3.63^b	42.43 ± 5.02^a	52.16 ± 2.02^{ab}	55.42 ± 3.59^b
T-AOC (U/mgprot)	1.76 ± 0.18^c	0.70 ± 0.11^a	1.14 ± 0.16^{ab}	1.46 ± 0.28^{bc}

ND：正常组；HFD：高脂模型组；NCU116-L：植物乳杆菌NCU116低剂量组；NCU116-H：植物乳杆菌NCU116高剂量组。结果以平均数±标准差表示（$n = 10$），同行不同上标字母表示组间具有显著性差异（$P < 0.05$）。ND: rats on the normal diet; HFD: rats on the high fat diet; NCU116-L: rats on the high fat diet +10^8 CFU/mL *L. plantarum* NCU116; NCU116-H: rats on the high fat diet +10^9 CFU/mL *L. plantarum* NCU116. Results are expressed as the means ± SEM ($n = 10$). Values within a row with different superscripts are significantly different ($P < 0.05$).

5.3.6 肝脏组织病理学观察

ND：正常组；HFD：高脂模型组；NCU116-L：植物乳杆菌NCU116低剂量组；NCU116-H：植物乳杆菌NCU116高剂量组。ND: rats on the normal diet; HFD: rats on the high fat diet; NCU116-L: rats on the high fat diet +10^8 CFU/mL *L. plantarum* NCU116: NCU116-H: rats on the high fat diet +10^9 CFU/mL *L. plantarum* NCU116.

如图5-6所示，正常组肝细胞结构清晰完整；高脂模型组肝脏有大量脂肪粒浸润，肝细胞呈空泡化坏死，脂肪肝病变明显；植物乳杆菌NCU116低剂量组与高剂量组肝细胞坏死程度减轻，脂肪粒浸润减少，说明植物乳杆菌NCU116能有效缓解脂肪肝病变程度。

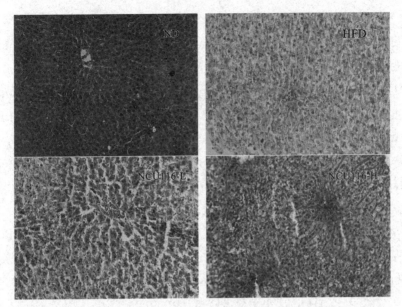

图 5-6 植物乳杆菌 NCU116 对高脂饮食大鼠肝脏病理学的影响（100X）

Figure 5-6 Effects of *L. plantarum* NCU116 on histology of liver in rats fed a high fat diet (100X).

5.3.7 脂代谢基因表达

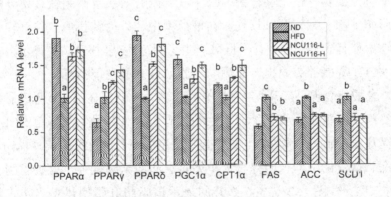

图 5-7 肝脏脂质分解、合成与脂肪酸氧化基因的表达

Figure 5-7 mRNA levels of lipolysis, lipogenesis and fatty acid oxidation genes in liver

ND：正常组；HFD：高脂模型组；NCU116-L：植物乳杆菌 NCU116 低剂量组；NCU116-H：植物乳杆菌 NCU116 高剂量组。结果以平均

数 ± 标准差表示（$n = 10$），图中不同上标字母表示组间具有显著性差异（$P < 0.05$）。ND: rats on the normal diet; HFD: rats on the high fat diet; NCU116-L: rats on the high fat diet $+10^8$ CFU/mL *L. plantarum* NCU116; NCU116-H: rats on the high fat diet $+10^9$ CFU/mL *L. plantarum* NCU116. ACC, Acetyl-coenzyme A carboxylase; CPT1α, Carnitine palmitoyltransferase-1α; FAS, Fatty acid synthetase; PGC1α, PPARγ coactivator-1α; PPAR, Peroxisome proliferator-activated receptor; SCD1, Coenzyme A desaturase 1. Data are expressed as the means ± SEM ($n = 10$). Values with different letters are significantly different ($P < 0.05$).

由图 5-7 可知，与正常组相比，高脂模型组显著改变了相关基因的表达水平（$P < 0.05$）。植物乳杆菌 NCU116 灌胃 5 周后，脂质代谢相关基因 PPARα、PPARγ、PPARδ 和 PGC1α 表达水平显著上调（$P < 0.05$）同时，FAS、ACC 和 SCD1 表达水平显著降低（$P < 0.05$）。

5.4 讨 论

本研究证实，植物乳杆菌 NCU116 可以有效缓解高脂饮食诱导的大鼠肝脏脂肪堆积、减轻肝脏损伤。在该模型中，谷丙转氨酶、谷草转氨酶和总胆红素都有显著升高，显示肝脏产生了明显病变。[189] 结果显示，植物乳杆菌 NCU116 能够有效提升非酒精性脂肪肝病肝功能指标。在一些动物模型中，乳酸菌表现出提升肝功能的作用。[190, 191] 研究显示，较低的谷丙转氨酶和谷草转氨酶活力水平意味着更好的肝功能。[192] 植物乳杆菌 NCU116 表现出提升肝功能的性质可能与该菌提升肠道屏障与降低肝脏病理损伤有关。

高脂饮食诱导的代谢综合征模型具有脂肪组织较大、内脏脂肪含量较多的特点。[193] 图 5-2 与表 5-2 共同提示该脂肪肝模型建立成功。研究证实，额外的营养补充可能导致脂肪组织肥大和胰岛素抵抗。前期研究显示，在高脂喂养的大鼠模型中，植物乳杆菌 NCU116 可以有效降低胰岛素抵抗、修复肝脏病理损伤、缓解脂肪细胞肥大症状。[185] 尽管其机制还没有完全阐明，但推测植物乳杆菌 NCU116 的抗脂肪肝作用可能与其干扰肝脏和脂肪组织中的代谢通路相关。[194]

动物经连续的高脂饮食可导致血清脂多糖（LPS）水平升高。[181] 肠道内革兰氏阴性菌代谢产生的 LPS 可以诱导促炎因子的分泌。促炎因子（如 TNF-α、IL-6）的分泌是机体对抑制体内致病菌入侵和定植的一种防御应答反应。[195]IL-10 一般被认为是一种抗炎因子，它由多种控制促炎反应的免疫调节分子组成。[196] 多个研究表明，益生菌能够通过促进促炎和抗炎因子的平衡来调节免疫失衡。[195] 促炎因子的分泌增加是非酒精性脂肪肝的早期症状。尤其要注意的是，TNF-α 和 IL-6 这两种细胞因子可能会促进代谢损伤。[183] 本研究中，植物乳杆菌 NCU116 可以明显降低 TNF-α、IL-6 与氧化应激水平，这可能是其缓解肝病理损伤的主要机制。[197]

肠道菌群的组成与能量摄入、体重增量和脂质代谢有关。连续高脂摄入可以改变肠道菌群组成，进而增加肠道通透性、激发炎症反应与诱导代谢紊乱。[60] 有研究表明，高脂饮食可以减少革兰氏阳性菌（如乳杆菌属和双歧杆菌属）的含量，增加革兰氏阴性菌（如肠杆菌属和拟杆菌属）的含量。[198]

改善非酒精性脂肪肝的肠道菌群失衡状态在研究中正逐渐被重视，这个方向涉及益生菌的多种生物学性质，如抑制致病菌、维持黏膜免疫与肠道屏障的完整性。[177] 益生菌作为肠道菌群的重要组成部分，具有促进脂质代谢的有益作用，其机制可能是益生菌能够影响胆固醇吸收与早期解离。[156] 前期研究发现，植物乳杆菌 NCU116 可以提高粪便中胆固醇和总胆汁酸含量，[185] 提示该菌可能可以吸收脂质来合成菌体细胞膜，同时促进粪便脂质排出，通过肠道菌群来改善机体的能量代谢。[156]

炎性因子提高、内脏脂肪积累都能够改变血脂水平，这些症状与胰岛素抵抗、血脂异常和心血管病有着密切关系。[169] 肝脏病理与脂肪酸组成分析发现，过多的肝脏脂肪堆积可能导致氧化应激和肝脏病理损伤，[156] 植物乳杆菌 NCU116 能够显著地降低肝脏脂肪积累，提示该菌具有抑制氧化应激和缓解肝病理损伤的能力。

肝脏脂质代谢主要通过脂代谢相关蛋白来调节，这些蛋白包括 β 氧化和脂肪生成蛋白。β 氧化是脂肪代谢的关键通路，它起到肝脏脂肪氧化指示剂的作用。[199, 200] 本研究为探讨植物乳杆菌 NCU116 对高脂饮食大鼠肝脂质代谢的可能机制，测定了脂肪氧化（PPARs，PGC1α 和 CPT1α）和脂肪生成（FAS，ACC 和 SCD1）的相关基因表达。结果显示，在两个植物乳杆菌 NCU116 剂量组中，脂肪合成、氧化基因的表达变化与炎性反应、血清脂多糖和 TNF-α 降低相伴发生。提示植物乳杆菌 NCU116 可以抑制高脂饮食诱导的炎性反应和肝脏的氧化应激。[164]另外，肝脏 mRNA 表达水平中，脂肪氧化基因（PPARs，PGC1α 和 CPT1α）显著升高，同时和脂肪生成基因（FAS，ACC 和 SCD1）显著降低。提示植物乳杆菌 NCU116 通过下调脂肪合成基因表达和上调脂肪氧化基因表达两条通路来降低肝脏脂肪积累。

5.5 本章小结

（1）植物乳杆菌 NCU116 在调节高脂饮食诱导的非酒精性脂肪肝病中，表现出抑制肝功能损伤、降低肝脏脂肪积累、改善脂肪肝的微观结构、促进肝损伤修复的作用。

（2）植物乳杆菌 NCU116 具有降低血清脂多糖和促炎因子，调节结肠菌群和脂代谢相关基因表达的作用。

（3）植物乳杆菌 NCU116 对肝脏脂代谢的调节可能是通过下调脂肪合成基因表达和上调脂肪氧化基因表达两条通路来实现的。

第6章 植物乳杆菌NCU116对高脂饮食大鼠血清代谢组学的影响

6.1 引 言

作为"组学"技术的一个重要分支，代谢组学技术也在随着生物科技的进步而不断发展，并在疾病的诊断和防治中发挥着越来越重要的作用。[201]代谢组学主要运用NMR、LC/MS、GC/MS等分析手段，通过定量检测血液、尿液等样本内的小分子化合物水平来研究机体代谢物的变化，探索代谢物和机体生理生化变化的关系（图6-1）。[202]随着社会生活水平的提高，人们的饮食和营养结构发生了巨大变化，高血糖、高血脂、高血压、脂肪肝、动脉粥样硬化等代谢综合征类疾病发病率日趋升高，这些疾病严重威胁着人类的身体健康。[203]

近年来，代谢组学的应用对多种代谢综合征的研究取得了一些成果。报道显示，这些疾病的发生和发展过程中，机体内代谢途径和代谢物水平在涉及能量代谢、葡萄糖代谢、脂质代谢与氨基酸代谢方面发生了明显变化。

在前期研究的基础上，利用高脂饲料喂养雄性SD大鼠形成的脂代谢紊乱动物模型，并灌胃不同剂量植物乳杆菌NCU116进行干预。利用UPLC-Q-TOF/MS检测该菌对高脂饮食大鼠代谢组学的影响，确定脂代谢紊乱可能的生物标记物，并阐述高脂饮食与代谢紊乱的关联性，为预防和治疗高脂饮食诱导的脂代谢紊乱提供新的思路和对策。

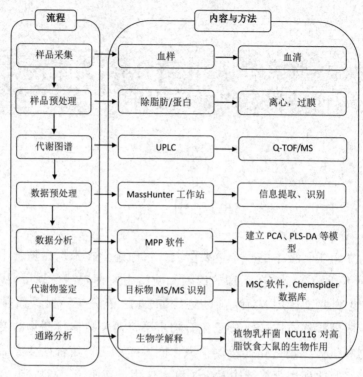

图 6-1　代谢组学研究过程

Figure 6-1 The procedures of metabolomics

6.2　实验部分

6.2.1　实验材料

6.2.1.1　实验动物

Sprague-Dawley（SD）大鼠，SPF 级，雄性，120～150 g，40 只，购自北京维通利华实验动物有限公司，许可证号：SCXK（京）2012-0001。动物饲料由南昌大学医学院实验动物中心提供。

饲养环境：温度 23 ℃ ±1 ℃，湿度 55% ±5%，光暗周期为 12 h/12 h 光照黑暗交替进行，实验前适应饲养 1 周，自由饮食饮水。

本实验动物操作获得南昌大学动物实验伦理委员会许可。

6.2.1.2 实验菌种

植物乳杆菌NCU116（南昌大学食品科学与技术国家重点实验室保藏）。

6.2.1.3 试剂耗材

甲酸（HPLC grade），美国Sigma公司；乙腈（HPLC grade），德国Merck公司；纯净水，广州屈臣氏公司。

6.2.2 实验设备

1290系列UPLC、6538系列Q-TOF-MS、ESI电喷雾离子源、Eclipse Plus C18色谱柱（3.5 μm, 2.1 mm i.d. × 150 mm）、Mass Hunter Qualitative Analysis软件、Mass Profiler Professional（MPP, B.02.00）软件，美国Agilent Technologies公司；代谢笼，苏州苏杭科技公司；冷冻离心机，美国Sigma公司。

6.2.3 实验方法

6.2.3.1 实验设计

大鼠实验环境下适应1周后，随机被分为正常组（ND）、高脂模型组（HFD）、植物乳杆菌NCU116低剂量组（NCU116-L，10^8 CFU/mL）和植物乳杆菌NCU116高剂量组（NCU116-H，10^9 CFU/mL），每组10只。给所有大鼠正常饮水，给予模型组、高剂量组和低剂量组高脂饲料；给予正常组正常饲料。按照10 mL/kg体重给高剂量组和低剂量组灌胃相应剂量的植物乳杆菌，高脂模型组和正常组灌胃同等剂量生理盐水，连续5周。

6.2.3.2 样品收集与前处理

灌胃结束后，心脏刺穿采血，分离血清。取200 μL血清加入800 μL乙腈-去离子水（4:1, v/v）中，涡旋混匀，离心，上清液过膜后，置于进样瓶中。

6.2.3.3 UPLC-Q-TOF/MS分析

安捷伦1290系列UPLC的分离参数为：进样量5 μL，Eclipse Plus C_{18}色谱柱（3.5 μm, 2.1 mm i.d. × 150 mm），柱温35 ℃，流动相A（0.1%甲酸水）和B（100%乙腈），流速0.3 mL/min，流动相洗脱梯度为：0~5 min 5%~20% B，5~10 min 20%~35% B，10~15 min 35%~98% B，15~16 min 98%~100% B and 16~18 min 100% B。

安捷伦 6538 Q-TOF/MS 质谱条件为：电喷雾离子源（ESI, electrospray ionization source）进行正（负）离子模式扫描，干燥气温度 325 ℃，干燥气流速 10 L/min，喷雾器压力 40 psig，裂解电压 120 V，锥孔电压 65 V，毛细管电压 3500 V，一级质谱扫描范围 50~1200 m/z，一级质谱扫描次数 2 spectra/s，二级质谱扫描次数 4 spectra/s。

6.2.3.4 数据处理与统计分析

一级质谱数据经 Agilent Mass Hunter Qualitative Analysis 软件分析，经分子特征查找，导出 CEF 格式文件。将各样品 CEF 文件经过 Mass Profiler Professional（MPP，B.02.00）软件分析。不同组代谢数据采用非配对 t 检验分析、建立主成分分析（PCA）和偏最小二乘法分析（PLS-DA）来研究各组代谢物的差异和分布规律。潜在生物标记物经 CAS、KEGG、HMP 和 METLIN 数据库分析确认。将分析得到的潜在生物标记物进行目标 UPLC-MS/MS 靶标分析，得到二级离子碎片，根据 MSC（molecular structure correlation）分析和 Chemspider 数据库综合分析，最终确认出高脂饮食大鼠代谢组学的生物标记物。利用 MPP 内置的 Pathway Analysis 工具对代谢物和代谢途径的相关性进行分析，发现植物乳杆菌 NCU116 对高脂饮食大鼠血液的生物标志物。

6.3 结果与分析

6.3.1 血清代谢物图谱分析

利用 UPLC-Q-TOF/MS 正、负离子模式全扫描可得血清总离子流色谱图，图 6-2 表示正离子检测模式（A）及负离子检测模式（B）下的离子流色谱图，显示正负离子模式都具有较好的分离度和离子丰度。另外，研究采用正离子（m/z 值为 121.0508 和 922.0097）和负离子（m/z 值为 112.39855 和 1034.9881）作为参比离子来确保正负离子扫描的质量准确性和重现性。

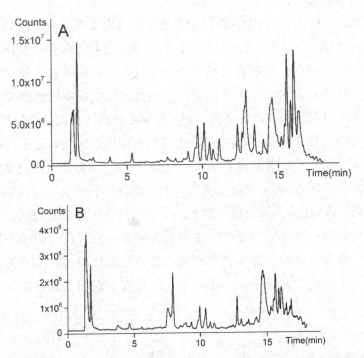

图6-2 高脂饮食大鼠血清样品总离子流图：（A）正离子模式，（B）负离子模式

Figure 6-2 (A) The positive and (B) negative total ion chromatogram of the serum sample from a high fat diet rat

6.3.2 代谢物多元变量分析

为了更好地体现植物乳杆菌NCU116在高脂饮食大鼠的代谢差异，首先对实验结果引入PCA分析。[204] 图6-3显示为PCA分析正离子（A）和负离子（B）的3D得分图，图中每个点代表一个样本，每个样本小分子代谢物的成分和浓度决定该样本在图中的位置，成分和浓度接近的样本在得分图上也比较靠近。结果显示，与正常组相比，高脂模型组分布位置较远，提示高脂模型组大鼠样本化合物组分和浓度与正常组有较大差异，高脂饮食大鼠体内生物化学过程有了明显改变。与高脂模型组相比，植物乳杆菌NCU116的低、高剂量组有向正常组靠近的趋势，其中高剂量组更为靠近正常组，说明植物乳杆菌NCU116高剂量组可以调节高脂饮食大鼠血液中小分子化合物的组分与含量。另外，正、负离子模式下PCA8种主成分含量经

MPP 软件分析分别为 70.59% 和 68.91%。

在 PCA 分析的基础上，PLS-DA 是一种可更清晰地表达组间差异的多变量的统计学方法。[205]PLS-DA 分析结果显示，4 组样本间差异更加明显（如图 6-4）。因此，植物乳杆菌 NCU116 改变了高脂饮食大鼠的血清代谢类型。植物乳杆菌 NCU116 高剂量组样本更为靠近正常组，表明植物乳杆菌 NCU116 高剂量组在改善高脂饮食大鼠血清代谢方面表现更为明显。

交叉验证用来评估 PLS-DA 建立模型的准确性。经验证矩阵模型分析可知，正、负离子扫描模式下准确率均为 100%。该验证矩阵模型数值在接近（或等于）100% 的情况下，表示该数学模型建立成功，可以相当准确地预测结果。[205]此外，该数据还显示样本组间具有很好的分离度；表示高脂饮食大鼠经植物乳杆菌 NCU116 灌胃后，血清小分子代谢物成分发生了较大变化。灌胃 5 周后，与高脂模型组比较植物乳杆菌 NCU116 组更为靠近正常组，提示植物乳杆菌 NCU116 在一定程度上影响了高脂饮食大鼠的代谢进程。

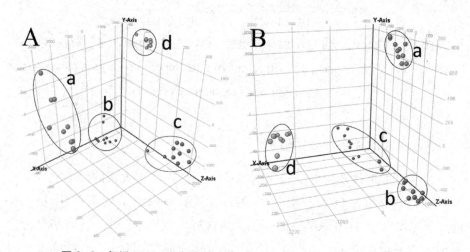

图 6-3　每组 3D PCA 得分图，（A）正离子模式，（B）负离子模式

Figure 6-3 Score plots in positive (A) and negative (B) ion mode from 3D PCA model classifying four groups.

a：正常组；b：高脂模型组；c：植物乳杆菌 NCU116 低剂量组（10^8 CFU/mL）；d：植物乳杆菌 NCU116 高剂量组（10^9 CFU/mL）。n=10。a: rats on the normal diet; b: rats on the high fat diet; c: rats on the high fat diet

$+10^8$ CFU/mL *L. plantarum* NCU116; d: rats on the high fat diet $+10^9$ CFU/mL *L. plantarum* NCU116. n = 10 in each group.

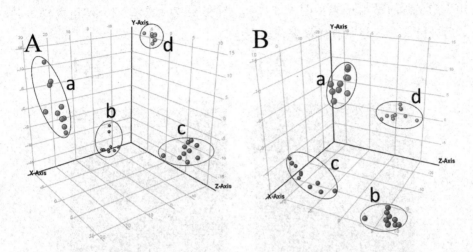

图 6-4 每组 3D PLS-DA 得分图, (A) 正离子模式, (B) 负离子模式

Figure 6-4 Score plots in positive (A) and negative (B) ion mode from 3D PLS-DA model classifying four groups.

a: 正常组; b: 高脂模型组; c: 植物乳杆菌 NCU116 低剂量组（10^8 CFU/mL）; d: 植物乳杆菌 NCU116 高剂量组（10^9 CFU/mL）。n=10。a: rats on the normal diet; b: rats on the high fat diet; c: rats on the high fat diet $+10^8$ CFU/mL *L. plantarum* NCU116; d: rats on the high fat diet $+10^9$ CFU/mL *L. plantarum* NCU116. n = 10 in each group.

a: 正常组; b: 高脂模型组; c: 植物乳杆菌 NCU116 低剂量组（10^8 CFU/mL）; d: 植物乳杆菌 NCU116 高剂量组（10^9 CFU/mL）。n=10。a: rats on the normal diet; b: rats on the high fat diet; c: rats on the high fat diet $+10^8$ CFU/mL *L. plantarum* NCU116; d: rats on the high fat diet $+10^9$ CFU/mL *L. plantarum* NCU116. n = 10 in each group.

聚类分析方法可对实验结果的同源性进行分析，即根据样本的相似性对它们进行分组，并通过聚类图来展示每个样本间的同源性关系。[206]对实验各组进行层次聚类分析（Hierarchical clustering analysis, HCA），发现在正离子模式下，植物乳杆菌 NCU116 低剂量组和高脂模型组同源性较高，再与植物乳杆菌 NCU116 高剂量组聚合，最后与正常组聚类，从而使所有样本聚合完毕（如图 6-5A1 所示）。在负离子模式下，植物乳杆菌 NCU116 低剂量组和高脂模型组同源性较高，再与正常组聚合，最后与植物乳杆菌 NCU116 高剂量组聚类（图 6-5B1）。图 6-5A2、B2 显示为各组内样本归一后进行聚合，与所有样本聚合趋势相同。

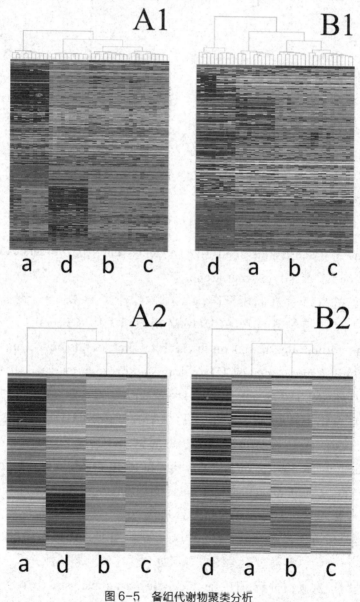

图 6-5 各组代谢物聚类分析

Figure 6-5 Hierarchical clustering analysis of metabolic profile data in four groups.

正离子聚合结果显示，植物乳杆菌 NCU116 高剂量组具有较好偏向正常组的趋势。负离子聚合结果显示，正常组在较长时间常规饲料喂养下，也发生了一定的向高脂模型组靠近的趋势，表明长时间的常规饲料自由摄食，也可能导致高脂血症的发生。此外，植物乳杆菌 NCU116 高剂量组远离高脂模型组的趋势更加明显。

6.3.3 代谢标志物的确证

本研究通过数据的系统检索和非配对 t 检验来匹配可能的生物标记物。利用筛选差异化合物流程，通过 Agilent METLIN Personal Metabolite Database 对标记物 ID 进行识别，匹配标记物 CAS 号，并利用 ID Browser 识别功能确定可能的生物标记物（表 6-1）。

通过 MPP 软件分析发现，经植物乳杆菌 NCU116 干预可明显改变一些代谢过程。利用 UPLC-Q-TOF/MS/MS 的二级离子碎片来鉴别潜在的生物标记物。通过碰撞电压为 10~40 eV 得到的 MS/MS 图谱碎片离子的信息导出为 CEF 格式文件，通过 MSC 软件和 Chemspider 数据库分析，最终鉴定本研究中的生物标记物。分析发现，正离子特征代谢标志物主要有亚精胺（Spermidine）、维生素 B5（Pantothenic Acid）、吲哚丙烯酸（Indoleacrylic acid）、吲哚（Indole）、5-羟吲哚乙醛（5-Hydroxy indole acetaldehyde）、甘氨胆酸（Glycocholic Acid）、胆氯素（Biliverdin IX）。负离子特征代谢标记物主要有羟基乙酸（Glycolic acid）、亮氨酸（L-Leucine）、2-苯基乙醇葡糖苷酸（2-Phenylethanol glucuronide）、牛磺胆酸（Taurocholic acid）、吲哚丙烯酸（Indoleacrylic acid）、甘氨胆酸（Glycocholic Acid）、2-花生酰基甘油（2-Arachidonoylglycerol）、孕二醇-3-葡萄糖苷酸（Pregnanediol-3-glucuronide）。

表6-1 血清潜在生物标记物的性质与各组变化趋势

Table 6-1 Metabolites selected as biomarkers characterized in serum profile and their change trends.

Ion modes	Compound Name	Molecular formula	Molecular weight	ND	HFD	NCU116-L	NCU116-H
Positive	Spermidine	$C_7H_{19}N_3$	145.245 895	8.212 38	−1.167	−4.222	−2.823
	Pantothenic Acid	$C_9H_{17}NO_5$	219.235 001	−9.423 2	3.052	3.004 7	3.366 5
	Indoleacrylic acid	$C_{11}H_9NO_2$	187.194 702	−16.938	5.6746	5.559 7	5.7039
	Indole	C_8H_7N	117.147 903	−11.578	4.3428	4.452 1	2.783 1
	5-Hydroxy indole acetaldehyde	$C_{10}H_9NO_2$	175.184 006	10.947 3	−3.649	−3.649	−3.649
	Glycocholic Acid	$C_{26}H_{43}NO_6$	465.622 711	−10.966	5.128 3	3.687 3	2.150 1
	Biliverdin IX	$C_{33}H_{34}N_4O_6$	582.646 301	2.044 09	5.688 8	−1.882	−5.851
Negative	Glycolic acid	$C_2H_4O_3$	76.051 399	4.087 903	−8.062 13	3.073 033	0.901 199
	L-Leucine	$C_6H_{13}NO_2$	131.172 897	−5.426 83	2.217 725	0.602 661	2.606 439
	2-Phenylethanol glucuronide	$C_{14}H_{18}O_7$	298.288 513	1.036 42	1.223 274	2.534 88	−4.794 58
	Taurocholic acid	$C_{26}H_{45}NO_7S$	515.703 003	1.717 932	1.513 942	1.961 599	−5.193 48
	Indoleacrylic acid	$C_{11}H_9NO_2$	187.194 702	7.549 991	−3.521 38	−6.214 24	2.185 628
	Glycocholic Acid	$C_{26}H_{43}NO_6$	465.622 711	−13.600 7	4.753 57	4.967 963	3.879 148
	2-Arachidonoylglycerol	$C_{23}H_{38}O_4$	378.545 41	−9.649 95	5.641 956	1.805 659	2.202 338
	Pregnanediol-3-glucuronide	$C_{27}H_{44}O_8$	496.633 514	−5.380 35	6.246 214	−1.1327 3	0.266 867

6.4 讨 论

表6-2和6-3显示，利用Pathway Analysis工具对已确证的小分子目标代谢物与代谢通路的相关性进行分析发现，目标代谢物参与机体内相应葡萄糖、脂肪酸、氨基酸、维生素、核苷酸等代谢途径。

植物乳杆菌 NCU116 可调节高脂饮食大鼠体内多胺类代谢途径。研究显示，亚精胺在 N_1- 乙酰转移酶作用下转变为 N_1- 乙酰基亚精胺，乙酰化的多胺类物质可被直接转运出细胞而降低体内多胺类物质的含量，进而影响羟脯胺赖氨酸在体内的代谢，来参与氨基酸代谢和衍生。[202]

5- 羟吲哚乙醛是 5- 羟色胺（5-HT，5-hydroxytryptamine）在单胺氧化酶作用下的衍生产物。[207] 5- 羟色胺参与糖尿病的色氨酸代谢调节，5- 羟吲哚乙醛的浓度与色氨酸代谢通路有关。[205] 色氨酸是人体的必需氨基酸，对体内多个代谢通路都具有重要的作用。

胆酸可以调控脂代谢吸收以及胆固醇的分解过程。牛磺胆酸（TCA）与甘氨胆酸（GCA）是胆酸代谢的产物，胆酸可能在脂肪酸生物合成、糖酵解、脂质与脂蛋白代谢中具有重要作用。[208] 研究报道，甘氨胆酸和胆氯素含量升高可诱发肝脏损伤。[209] 另外，亮氨酸在胆酸代谢和葡萄糖平衡通路中具有重要作用。定量分析发现，植物乳杆菌 NCU116 高剂量组具有较好降低牛磺胆酸、甘氨胆酸和胆氯素的性能。前期研究发现，植物乳杆菌 NCU116 表现出缓解脂代谢紊乱的特性，与本研究该菌能够影响血清 TCA 和 GCA 浓度结果具有一定的相关性。[185]

维生素 B5 是 B 族维生素中重要的一种，在抗体合成、能量代谢方面具有重要作用。本研究发现，高脂饮食能够增加维生素 B5 的含量，但植物乳杆菌 NCU116 缓解这一现象。羟基乙酸、2- 花生酰基甘油、孕二醇 -3- 葡萄糖苷酸、2- 苯基乙醇葡糖苷酸、吲哚和吲哚丙烯酸它们在色氨酸、天冬酰胺等氨基酸代谢方面具有重要作用。

本实验利用 UPLC-Q-TOF/MS 来研究植物乳杆菌 NCU116 对高脂饮食大鼠在生物学层面的影响。不同生物标记物的浓度经植物乳杆菌 NCU116 干预 5 周后发生变化，结果显示植物乳杆菌 NCU116 在吲哚丙烯酸、甘氨胆酸、胆氯素、羟基乙酸、牛磺胆酸、2- 花生酰基甘油和孕二醇 -3- 葡萄糖苷酸等生物标记物上具有较好的调节作用。另外，植物乳杆菌 NCU116 组两个剂量组含有不同浓度的植物乳杆菌 NCU116，并且在统计分析时高剂量组和正常组表现得更为接近（图 6-2 和图 6-3）。因此，表明植物乳杆菌 NCU116 高剂量组表现出的缓解高脂饮食诱导脂肪肝和高脂血症症状的功能与植物乳杆菌 NCU116 高剂量补充有关。

表6-2 正离子模式下潜在生物标记物鉴定及相关代谢通路

Table 6-2 Potential biomarkers identified in positive ion mode and the related pathways.

Retention time _ M+	MS [M+H]+	Proposed compound	Proposed structure	MS/MS	Related pathway
1.51	146.16504	Spermidine		98.0948 $C_6H_{12}N$ 84.0814 $C_5H_{10}N$ 72.0818 $C_4H_{10}N$ 56.0501 C_3H_6N	Urea cycle and metabolism of amino groups Nucleotide Metabolism Phase 1 – Functionalization of compounds Metabolism of amino acids and derivatives
4.35-25	220.11797	Pantothenic Acid		131.1264 $C_7H_{17}NO$ 115.0349 $C_5H_7O_3$ 98.0346 $C_5H_6O_2$ 85.0607 C_5H_9O	Metabolism of water-soluble vitamins and cofactors
5.465	188.07065	Indoleacrylic acid		144.0803 $C_{10}H_{10}N$ 132.0794 $C_9H_{10}N$ 118.0650 C_8H_8N 55.0172 C_3H_3O	Metabolism of amino acids and derivatives

续表

Retention time_M+	MS [M+H]+	Proposed compound	MS/MS	Proposed structure	Related pathway
5.466 21	118.065 12	Indole	103.054 4 C_8H_7 77.039 6 C_6H_5 58.065 4 C_3H_8N 55.054 5 C_4H_7		Tryptophan metabolism Phase 1 - Functionalization of compounds
11.165	176.070 53	5-Hydroxyindoleacetaldehyde	130.065 5 C_9H_8N 128.061 9 $C_{10}H_8$ 112.048 6 $C_6H_8O_2$ 92.053 3 C_6H_6N 77.037 4 C_6H_5 58.073 3 C_4H_{10}		Tryptophan metabolism Phase 1 - Functionalization of compounds
12.791 1	466.316 38	Glycocholic Acid	430.298 1 $C_{26}H_{40}NO_4$ 337.257 4 $C_{20}H_{35}NO_3$ 231.173 2 $C_{16}H_{23}O$ 209.117 6 $C_{12}H_{17}O_3$ 144.065 1 $C_6H_{10}NO_3$ 90.055 3 $C_3H_8NO_2$		Fatty Acid Biosynthesis, Glycolysis and Gluconeogenesis, Metabolism of lipids and lipoproteins, Glucose Homeostasis Drug Induction of Bile Acid Pathway

续表

Retention time_M+	MS [M+H]+	Proposed compound	MS/MS	Proposed structure	Related pathway
14.963	583.25540	Biliverdin IX	524.2452 $C_{31}H_{32}N_4O_4$ 445.2240 $C_{26}H_{29}N_4O_3$ 417.1448 $C_{24}H_{21}N_2O_5$ 387.1591 $C_{22}H_{19}N_4$ 240.1001 $C_{15}H_{14}NO_2$ 233.1149 $C_{14}H_{17}O_3$ 86.0959 $C_5H_{12}N$		Metabolism of porphyrins

表6-3 负离子模式下潜在生物标记物鉴定及相关代谢通路

Table 6-3 Potential biomarkers identified in positive ion mode and the related pathways.

Retention time_M−	MS [M−H]−	Proposed compound	MS/MS	Proposed structure	Related pathway
.46091	75.00909	Glycolic acid	73.9968 $C_2H_2O_3$		Urea cycle and metabolism of amino groups

续表

Retention time_M-	MS [M-H]-	Proposed compound	MS/MS	Proposed structure	Related pathway
2.655 67	130.087 31	L-Leucine	88.041 8 $C_3H_6NO_2$ 87.045 8 $C_4H_7O_2$ 85.030 6 $C_4H_5O_2$ 68.997 0 C_3HO_2		Drug Induction of Bile Acid Pathway Glucose Homeostasis
10.721 2	297.097 90	2-Phenylethanol glucuronide	218.037 4 $C_8H_{10}O_7$ 125.098 9 $C_8H_{13}O$ 101.025 2 $C_4H_5O_3$ 85.030 1 $C_4H_5O_2$		Asparagine N-linked glycosylation Transport of inorganic cations-anions and amino acids oligopeptides Nicotine metabolism
12.162 8	514.284 70	Taurocholic acid	370.312 6 $C_{25}H_{40}NO$ 342.219 4 $C_{19}H_{34}O_3S$ 241.223 1 $C_{15}H_{29}O_2$ 124.008 2 $C_2H_6NO_3S$ 79.957 2 O_3S		Glucose Homeostasis

续表

Retention time_M-	MS [M-H]-	Proposed compound	MS/MS	Proposed structure	Related pathway
12.537 6	186.055 74	Indoleacrylic acid	142.066 3 $C_{10}H_8N$ 115.040 3 C_8H_5N 103.040 8 C_7H_5 89.024 2 C_6H_3N 85.030 0 $C_4H_5O_2$ 75.009 8 C_5HN		Urea cycle and metabolism of amino groups
12 849 3	464.301 97	Glycocholic Acid	419.237 9 $C_{24}H_{35}O_6$ 330.223 6 $C_{21}H_{30}O_3$ 233.160 8 $C_{11}H_{23}NO_4$ 227.167 3 $C_{16}H_{21}N$ 101.080 3 $C_5H_{11}NO$ 74.024 2 $C_2H_4NO_2$		Nucleotide Metabolism Glucose Homeostasis
13.836 3	377.269 60	2-Arachidonoylglycerol	275.166 2 $C_{17}H_{23}O_3$ 231.175 7 $C_7H_{23}O$ 221.154 5 $C_{14}H_{21}O_2$ 113.024 1 $C_5H_5O_3$ 79.060 3 C_6H_7 68.995 8 C_3HO_2		GPCR downstream signaling Metabolism of amino acids and derivatives

续 表

Retention time_M-	MS [M-H]-	Proposed compound	MS/MS	Proposed structure	Related pathway
14.338 7	495.296 08	Pregnanediol-3-glucuronide	405.268 7 $C_{24}H_{37}O_5$ 391.288 7 $C_{24}H_{39}O_4$ 375.182 1 $C_{21}H_{27}O_6$ 311.183 8 $C_{17}H_{27}O_5$ 140.011 7 $C_6H_4O_4$ 103.039 2 $C_4H_7O_3$ 89.025 1 $C_3H_5O_3$		Asparagine N-linked glycosylation Transport of inorganic cations-anions and amino acids-oligopeptides

6.5 本章小结

本章通过高脂喂养诱导来建立高脂血症大鼠模型，利用 UPLC-Q-TOF/MS 来研究植物乳杆菌 NCU116 对血清代谢组学的影响。通过非配对 t 检验、PCA、PLS-DA 和 HCA 分析可知，高脂饮食诱导的高脂血症大鼠血清中亚精胺（Spermidine）、维生素 B5（Pantothenic Acid）、吲哚丙烯酸（Indoleacrylic acid）、吲哚（Indole）、5-羟吲哚乙醛（5-Hydroxy indole acetaldehyde）、甘氨胆酸（Glycocholic Acid）、胆氯素（Biliverdin IX）、羟基乙酸（Glycolic acid）、亮氨酸（L-Leucine）、2-苯基乙醇葡糖苷酸（2-Phenylethanol glucuronide）、牛磺胆酸（Taurocholic acid）、2-花生酰基甘油（2-Arachidonoylglycerol）、孕二醇-3-葡萄糖苷酸（Pregnanediol-3-glucuronide）代谢均有影响。但是，植物乳杆菌 NCU116 仅在吲哚丙烯酸、甘氨胆酸、胆氯素、羟基乙酸、牛磺胆酸、2-花生酰基甘油和孕二醇-3-葡萄糖苷酸等生物标记物上发挥较好的调节作用，并通过这些生物标记物来调节机体内脂质、葡萄糖和脂蛋白等代谢途径来发挥抑制高脂血症作用。另外，植物乳杆菌 NCU116 表现出的缓解高脂血症症状的功能与植物乳杆菌 NCU116 高剂量补充有关。

第7章 植物乳杆菌NCU116及发酵胡萝卜汁对小鼠结肠炎的缓解作用

7.1 引 言

  炎症性肠病（Inflammatory bowel disease，IBD）是一种发病机制至今还没有完全阐明的慢性非特异性肠道疾病。其主要包括克罗恩病（Crohn's disease，CD）、溃疡性结肠炎（Ulcerative Colitis，UC）和结肠袋炎（Pouchitis）。[210, 211] 目前一般认为，环境、遗传、免疫和肠道菌群等多种因素是影响炎症性肠病的关键因子。[212]

  药物治疗是目前结肠炎治疗的常用手段，但其过高的价格和不可预期的副作用限制了药物的适用性和功效。[213] 鉴于以上忧虑，研究安全、实惠与副作用较低的预防和治疗结肠炎的替代方案显得尤为重要。如前面章节所述，益生菌在改善人体肠道健康、促进肠道微生态平衡、调节血糖血脂和免疫平衡方面具有重要作用。近年来，益生菌在结肠炎发病和治疗过程中的作用越来越受到重视。[214]

  本章拟在植物乳杆菌NCU116及其发酵胡萝卜汁摄入对三硝基苯磺酸（2，4，6-trinitrobenzenesulfonic acid, TNBS）诱导结肠炎小鼠的缓解作用方面进行研究。通过评价体重、结肠表观与微观形态、细胞因子、氧化应激和短链脂肪酸等方面对结肠炎损伤的修复进行评估，并得出植物乳杆菌NCU116及其发酵胡萝卜汁对结肠炎小鼠可能的作用机制。

7.2 实验部分

7.2.1 实验材料

7.2.1.1 实验动物

Balb/c小鼠，雄性，20~22g，50只，购自北京维通利华实验动物有限公司，许可证号：SCXK（京）2012-0001。动物饲料由南昌大学医学院实验动物中心提供。

饲养环境：温度23℃±1℃，湿度55%±5%，光暗周期为12 h/12 h光照黑暗交替进行，实验前适应饲养1周，自由饮食饮水。

本实验动物操作获得南昌大学动物实验伦理委员会许可。

7.2.1.2 实验菌种

植物乳杆菌NCU116菌种（南昌大学食品科学与技术国家重点实验室保藏）。

7.2.1.3 试剂耗材

胡萝卜，购于南昌当地市场；三硝基苯磺酸（TNBS），美国Sigma公司；超氧化物歧化酶（SOD）、谷胱甘肽过氧化物酶（GSH-Px）、过氧化氢酶（CAT）、丙二醛（MDA）和乳酸试剂盒，南京建成生物工程研究所；IFN-γ、TNF-α、IL-6、IL-10和IL-12酶联免疫试剂盒（ELISA法），武汉博士德生物工程有限公司；乙酸、丙酸、正丁酸、异丁酸标准品，上海阿拉丁试剂公司；10%中性福尔马林固定液，南昌市雨露实验器材有限公司；苏木素染液、伊红染液，碧云天生物技术公司；其他试剂均为国产分析纯。

7.2.2 实验设备

SHP-150生化培养箱，上海森信实验仪器有限公司；JJ-CJ-ZFD洁净工作台，吴江市净化设备总厂；TGL-16G-A离心机，上海安亭科学仪器厂；YP10002马头牌电子天平，上海光正医疗仪器有限公司；AL104型电子天平，上海梅特勒-托利多仪器公司；6890N气相色谱仪，美国Agilent Technologies公司；Varioskan Flash全波长多功能酶标仪，美国Thermo

Scientific 公司；CX31 电子显微镜，日本 Olympus 公司；KD2258 生物组织切片机、KD-T 电脑生物组织摊烤片机、D-BM 生物组织包埋机、KD-BL 包埋机冷冻台、KD-TS3D 生物组织脱水机，浙江科迪仪器设备有限公司。

7.2.3 实验方法

7.2.3.1 结肠炎模型的制备

小鼠适应实验环境后，禁食12h，用3.5F导管从肛门插入肠道约5.5 cm处，灌注三硝基苯磺酸（TNBS）50%乙醇溶液 100 μL（含 2 mg TNBS 的 50% 乙醇溶液），之后将小鼠倒置60s，防止液体流出，建立溃疡性损伤结肠炎模型。将造模小鼠随机分为 4 组：分别为（B）结肠炎模型组（IBD），灌胃生理盐水；（C）植物乳杆菌 NCU116 组（NCU），灌胃 NCU116 植物乳杆菌 NCU116（10^9 CFU/mL）生理盐水悬浮液；（D）发酵胡萝卜汁组（FCJ），灌胃植物乳杆菌 NCU116 发酵胡萝卜汁（含菌量 10^9 CFU/mL）；（E）胡萝卜汁组（NFCJ），灌胃未发酵胡萝卜汁。另设（A）正常组（Normal），灌注同等剂量50%乙醇溶液，灌胃生理盐水。每组 10 只动物，每天按 10 mL/kg 连续给药 5 周。动物自由摄食饮水。

7.2.3.2 发酵胡萝卜汁的制作工艺

取新鲜胡萝卜打碎，巴氏杀菌，接种活化的植物乳杆菌 NCU116 菌种，37 ℃ 发酵24h。发酵后菌落计数得发酵胡萝卜汁中菌含量约为 10^9 CFU/mL。另取未接种植物乳杆菌 NCU116 的胡萝卜汁在同等条件下发酵，作为对照组。

7.2.3.3 发酵胡萝汁与粪便短链脂肪酸测定

参照 2.2.3.4 步骤进行。

7.2.3.4 体重与结肠长度变化

实验过程中，记录小鼠体重变化。灌胃结束后，小鼠眼眶取血，脱椎处死，取结肠，观察各组小鼠结肠的形态改变并测量结肠长度。

7.2.3.5 血清氧化应激水平测定

参照 2.2.3.7 步骤进行。

7.2.3.6 血清细胞因子测定

参照 2.2.3.6 步骤进行。

7.2.3.7 结肠病理学观察

参照 3.2.3.7 步骤进行。

7.2.3.8 统计学分析

各实验组数据以平均数 ± 标准差 ($\bar{x} \pm s$) 表示，采用 SPSS 17.0 软件进行数据统计分析，Duncan's 多重范围检验。$P < 0.05$ 表示组间具有显著性差异，具有统计学意义。

7.3 结果与分析

7.3.1 发酵胡萝卜汁中短链脂肪酸分析

如表 7-1 所示，与未发酵胡萝卜汁相比，经植物乳杆菌 NCU116 发酵胡萝卜汁具有更高浓度的乙酸、丙酸。其中，未发酵胡萝卜汁中未检出丁酸，而发酵胡萝卜汁中含有 0.25 mmol/L 的丁酸。

表7-1 发酵胡萝卜汁中短链脂肪酸水平（mmol/L）

Table 7-1 Short chain fatty acids (SCFA) in fermented carrot juices (mmol/L)

Parameters	NFCJ	FCJ
Acetic acid	8.99 ± 0.41	18.58 ± 0.66
Propionic acid	0.31 ± 0.02	0.47 ± 0.03
Butyric acid	nd	0.25 ± 0.02
SCFA	9.30 ± 0.42	19.29 ± 0.67

FCJ：植物乳杆菌 NCU116 发酵胡萝卜汁；NFCJ：未发酵胡萝卜汁。nd：未检出。总短链脂肪酸 = 乙酸 + 丙酸 + 丁酸。结果以平均数 ± 标准差表示（$n = 6$）。FCJ: Fermented carrot juice with L. plantarum NCU116, NFCJ: non-fermented carrot juice. "nd" means not detected. SCFA = Acetic acid + Propionic acid + Butyric acid. Results are expressed as the means ± SEM ($n = 6$).

7.3.2 体重变化

如表7-2所示,在模型建立前(第0周)各组体重基本无差异。模型建立并给药1周后,正常组体重有所增长,其他各组体重明显降低($P < 0.05$)。2周后,每组体重均有所增长。3周后,植物乳杆菌NCU116及发酵胡萝卜汁组体重显著高于结肠炎模型组($P < 0.05$)。实验进行5周时,植物乳杆菌NCU116及发酵胡萝卜汁体重较结肠炎模型组有所升高,但结果不具有显著性差异。

表7-2 植物乳杆菌NCU116及发酵胡萝卜汁对结肠炎小鼠体重的影响

Table 7-2 Effect of NCU and FCJ on body weight of IBD mice.

Body weight	0th week	1st week	2nd week	3rd week	4th week	5th week
Normal	24.45 ± 0.43	25.92 ± 0.77[b]	28.05 ± 0.76[b]	28.54 ± 0.64[c]	29.34 ± 0.61[b]	30.01 ± 0.50[b]
IBD	24.10 ± 0.51	21.84 ± 0.50[a]	22.74 ± 0.48[a]	23.98 ± 0.59[a]	26.46 ± 0.72[a]	27.30 ± 0.73[a]
NCU	24.37 ± 0.15	22.04 ± 0.53[a]	24.77 ± 0.76[a]	26.41 ± 0.53[bc]	27.48 ± 0.47[ab]	28.48 ± 0.47[ab]
FCJ	24.70 ± 0.71	22.59 ± 0.27[a]	25.18 ± 1.20[a]	26.49 ± 0.91[bc]	28.14 ± 0.77[ab]	29.01 ± 0.96[ab]
NFCJ	24.20 ± 0.41	22.05 ± 0.77[a]	23.29 ± 1.21[a]	24.66 ± 1.09[ab]	26.37 ± 0.94[a]	26.82 ± 0.99[a]

Normal:正常组,灌胃生理盐水;IBD:结肠炎模型组,灌胃生理盐水;NCU:植物乳杆菌NCU116组,灌胃NCU116植物乳杆菌NCU116(10^9 CFU/mL)生理盐水悬浮液;FCJ:发酵胡萝卜汁组,灌胃植物乳杆菌NCU116发酵胡萝卜汁(含菌量10^9 CFU/mL);NFCJ:(未发酵)胡萝卜汁组,灌胃未发酵胡萝卜汁。结果以平均数 ± 标准差表示($n = 10$),同列不同上标字母表示组间具有显著性差异($P < 0.05$)。Normal, non-IBD+saline; IBD, IBD + saline; NCU: IBD + 10^9 CFU/mL $L.$ $plantarum$ NCU116; FCJ: IBD + fermented carrot juice with $L.$ $plantarum$ NCU116; NFCJ: IBD + non-fermented carrot juice. Data are expressed as the means ± SEM ($n = 10$). Values within a column with different superscripts are significantly different ($P < 0.05$).

7.3.3 结肠重量与长度变化

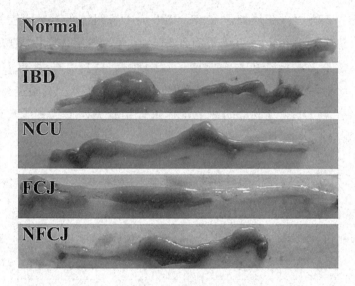

图 7-1　结肠炎模型小鼠结肠组织表观变化

Figure 7-1 Macroscopic appearances of the colon from IBD mouse model

Normal：正常组，灌胃生理盐水；IBD：结肠炎模型组，灌胃生理盐水；NCU：植物乳杆菌 NCU116 组，灌胃 NCU116 植物乳杆菌 NCU116（10^9 CFU/mL）生理盐水悬浮液；FCJ：发酵胡萝卜汁组，灌胃植物乳杆菌 NCU116 发酵胡萝卜汁（含菌量 10^9 CFU/mL）；NFCJ：（未发酵）胡萝卜汁组，灌胃未发酵胡萝卜汁。Normal, non-IBD+ saline; IBD, IBD + saline; NCU: IBD + 10^9 CFU/mL *L. plantarum* NCU116; FCJ: IBD + fermented carrot juice with *L. plantarum* NCU116; NFCJ: IBD + non-fermented carrot juice.

从结肠表观观察可知，与正常组相比，结肠炎模型组结肠表现出肠道扭曲、变形，结肠与其他相邻组织有粘连，结肠中段膨大，并伴有糜烂、溃疡现象。植物乳杆菌 NCU116 及发酵胡萝卜汁组上述症状明显好转（图 7-1）。称量后可知，各组结肠重量并无显著性差异，但结肠炎模型组和胡萝卜汁组结肠重量高于其他各组。在结肠长度方面，与正常组相比，结肠炎模型组结肠长度显著降低（$P < 0.05$）；植物乳杆菌 NCU116 及发酵胡萝卜汁组较结肠炎模型组有所增长，但结果不具有显著性差异（图 7-2）。

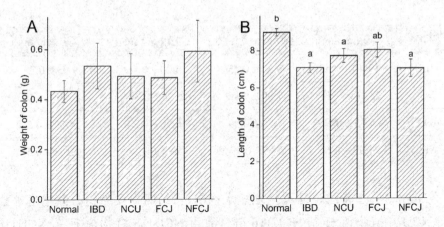

图 7-2 植物乳杆菌 NCU116 及发酵胡萝卜汁对结肠炎小鼠结肠重量和长度的影响

Figure 7-2 Effect of NCU and FCJ treatment on length and weight of colon in IBD mice.

Normal：正常组，灌胃生理盐水；IBD：结肠炎模型组，灌胃生理盐水；NCU：植物乳杆菌 NCU116 组，灌胃 NCU116 植物乳杆菌 NCU116（10^9 CFU/mL）生理盐水悬浮液；FCJ：发酵胡萝卜汁组，灌胃植物乳杆菌 NCU116 发酵胡萝卜汁（含菌量 10^9 CFU/mL）；NFCJ：（未发酵）胡萝卜汁组，灌胃未发酵胡萝卜汁。结果以平均数 ± 标准差表示（$n = 10$），不同上标字母表示组间具有显著性差异（$P < 0.05$）。Normal, non-IBD+ saline; IBD, IBD + saline; NCU: IBD + 10^9 CFU/mL *L. plantarum* NCU116; FCJ: IBD + fermented carrot juice with *L. plantarum* NCU116; NFCJ: IBD + non-fermented carrot juice. Data are expressed as the means ± SEM ($n = 10$). Values with different superscripts are significantly different ($P < 0.05$).

7.3.4 结肠粪便短链脂肪酸变化

表7-3 植物乳杆菌NCU116及发酵胡萝卜汁对结肠炎小鼠粪便短链脂肪酸含量的影响（μmol/g）

Table 7-3 Changes in short chain fatty acids (SCFA) in feces of five groups of mice (μmol/g).

SCFA	Acetic acid	Propionic acid	i-Butyric acid	n-Butyric acid
Normal	25.61 ± 0.64[d]	7.93 ± 0.74[b]	6.70 ± 0.50[b]	15.81 ± 1.04[b]

续表

SCFA	Acetic acid	Propionic acid	i-Butyric acid	n-Butyric acid
IBD	19.40 ± 0.50^a	5.38 ± 0.25^a	5.20 ± 0.46^a	9.95 ± 0.73^a
NCU	23.36 ± 0.71^{bc}	7.99 ± 0.66^b	6.28 ± 0.36^{ab}	14.78 ± 1.01^b
FCJ	24.13 ± 0.67^{cd}	7.76 ± 0.99^b	6.90 ± 0.24^b	14.59 ± 0.56^b
NFCJ	21.97 ± 0.60^b	6.95 ± 0.47^{ab}	6.25 ± 0.28^{ab}	11.01 ± 0.50^a

Normal：正常组，灌胃生理盐水；IBD：结肠炎模型组，灌胃生理盐水；NCU：植物乳杆菌NCU116组，灌胃NCU116植物乳杆菌NCU116（10^9 CFU/mL）生理盐水悬浮液；FCJ：发酵胡萝卜汁组，灌胃植物乳杆菌NCU116发酵胡萝卜汁（含菌量10^9 CFU/mL）；NFCJ：（未发酵）胡萝卜汁组，灌胃未发酵胡萝卜汁。结果以平均数 ± 标准差表示（$n = 10$），同列不同上标字母表示组间具有显著性差异（$P < 0.05$）。Normal, non-IBD+ saline; IBD, IBD + saline; NCU: IBD + 10^9 CFU/mL $L.\ plantarum$ NCU116; FCJ: IBD + fermented carrot juice with $L.\ plantarum$ NCU116; NFCJ: IBD + non-fermented carrot juice. Data are expressed as the means ± SEM ($n = 10$). Values within a column with different superscripts are significantly different ($P < 0.05$).

如表7-3所示，与正常组相比，结肠炎模型组乙酸、丙酸、正丁酸、异丁酸含量均有显著降低（$P < 0.05$）。灌胃植物乳杆菌NCU116及发酵胡萝卜汁5周后，乙酸、丙酸、正丁酸含量均有显著升高（$P < 0.05$）。未发酵胡萝卜汁能够升高各种短链脂肪酸水平，但仅在乙酸水平上具有显著性差异（$P < 0.05$）。

7.3.5 血清细胞因子水平

如表7-4所示，与正常组相比，结肠炎模型组IL-6、IL-12、TNF-α和IFN-γ水平显著升高，IL-10水平显著降低（$P < 0.05$）。与结肠炎模型组相比，植物乳杆菌NCU116及发酵胡萝卜汁组IL-6、IL-12、TNF-α和IFN-γ水平显著降低，IL-10水平显著升高（$P < 0.05$）。此外，未发酵胡萝卜汁组在IL-6、IL-10、IL-12和TNF-α指标上，也表现出一定的缓解作用。

表7-4 植物乳杆菌NCU116及发酵胡萝卜汁对结肠炎小鼠细胞因子的影响（pg/mL）

Table 7-4 Effect of NCU and FCJ treatment on cytokines in IBD mice (pg/mL)

Cytokines	IL-6	IL-10	IL-12	TNF-α	IFN-γ
Normal	146.41 ± 12.49^a	130.65 ± 4.55^d	71.24 ± 5.77^a	66.57 ± 3.45^a	33.83 ± 2.72^a
IBD	304.62 ± 13.30^c	38.85 ± 4.86^a	189.06 ± 5.84^d	126.07 ± 2.82^c	102.62 ± 3.00^d
NCU	150.69 ± 7.75^a	88.17 ± 5.34^c	104.58 ± 4.58^b	69.33 ± 2.25^a	75.00 ± 3.21^c
FCJ	147.64 ± 12.56^a	122.29 ± 6.72^d	94.66 ± 6.13^b	63.39 ± 3.09^a	54.20 ± 2.96^b
NFCJ	223.47 ± 12.12^b	57.91 ± 3.68^b	161.06 ± 4.25^c	104.62 ± 2.40^b	101.54 ± 3.52^d

Normal：正常组，灌胃生理盐水；IBD：结肠炎模型组，灌胃生理盐水；NCU：植物乳杆菌NCU116组，灌胃NCU116植物乳杆菌NCU116（10^9 CFU/mL）生理盐水悬浮液；FCJ：发酵胡萝卜汁组，灌胃植物乳杆菌NCU116发酵胡萝卜汁（含菌量10^9 CFU/mL）；NFCJ：（未发酵）胡萝卜汁组，灌胃未发酵胡萝卜汁。结果以平均数 ± 标准差表示（$n = 10$），同列不同上标字母表示组间具有显著性差异（$P < 0.05$）。Normal, non-IBD+ saline; IBD, IBD + saline; NCU: IBD + 10^9 CFU/mL L. plantarum NCU116; FCJ: IBD + fermented carrot juice with L. plantarum NCU116; NFCJ: IBD + non-fermented carrot juice. Data are expressed as the means ± SEM ($n = 10$). Values within a column with different superscripts are significantly different ($P < 0.05$).

7.3.6 血清氧化应激水平

表7-5 植物乳杆菌NCU116及发酵胡萝卜汁对结肠炎小鼠氧化应激的影响

Table 7-5 Effect of NCU and FCJ treatment on oxidative stress in IBD mice

Oxidative stress	SOD (U/mL)	GSH-Px (U/mL)	MDA (nmol/mL)	CAT (U/mL)
Normal	160.35 ± 3.55^d	1663.79 ± 24.32^b	9.86 ± 0.35^a	10.74 ± 0.18^c
IBD	97.60 ± 3.41^a	1299.85 ± 23.95^a	16.51 ± 0.33^d	5.38 ± 0.25^a
NCU	150.60 ± 3.48^d	1627.34 ± 21.23^b	12.25 ± 0.34^b	9.89 ± 0.30^b

Oxidative stress	SOD (U/mL)	GSH–Px (U/mL)	MDA (nmol/mL)	CAT (U/mL)
FCJ	135.28 ± 4.84[c]	1661.49 ± 22.22[b]	10.96 ± 0.36[a]	9.29 ± 0.27[b]
NFCJ	116.65 ± 4.60[b]	1253.00 ± 21.17[a]	15.11 ± 0.53[c]	5.87 ± 0.25[a]

Normal：正常组，灌胃生理盐水；IBD：结肠炎模型组，灌胃生理盐水；NCU：植物乳杆菌 NCU116 组，灌胃 NCU116 植物乳杆菌 NCU116（10^9 CFU/mL）生理盐水悬浮液；FCJ：发酵胡萝卜汁组，灌胃植物乳杆菌 NCU116 发酵胡萝卜汁（含菌量 10^9 CFU/mL）；NFCJ：（未发酵）胡萝卜汁组，灌胃未发酵胡萝卜汁。结果以平均数 ± 标准差表示（$n = 10$），同列不同上标字母表示组间具有显著性差异（$P < 0.05$）。Normal, non-IBD+ saline; IBD, IBD + saline; NCU: IBD + 10^9 CFU/mL L. plantarum NCU116; FCJ: IBD + fermented carrot juice with L. plantarum NCU116; NFCJ: IBD + non-fermented carrot juice. Data are expressed as the means ± SEM ($n = 10$). Values within a column with different superscripts are significantly different ($P < 0.05$).

如表 7-5 所示，与正常组相比，结肠炎模型组 SOD、GSH-Px 和 CAT 水平显著降低，MDA 水平显著升高（$P < 0.05$）。与结肠炎模型组相比，植物乳杆菌 NCU116 及发酵胡萝卜汁组 SOD、GSH-Px 和 CAT 水平显著升高，MDA 水平显著降低。未发酵胡萝卜汁组在 SOD 和 MDA 损伤水平上表现出一定的缓解作用。

7.3.7 结肠病理变化

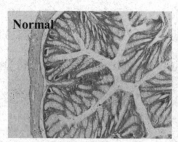
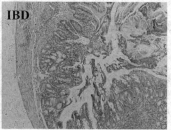

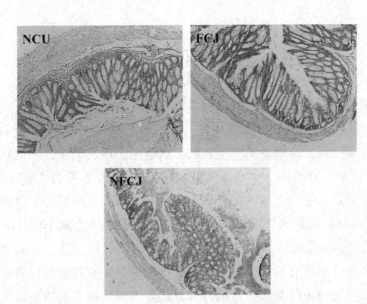

图7-3 植物乳杆菌NCU116及发酵胡萝卜汁对结肠炎小鼠结肠病理损伤的修复作用

Figure 7-3 Effect of NCU and FCJ treatment on histology of colon in IBD mice (100X)

Normal：正常组，灌胃生理盐水；IBD：结肠炎模型组，灌胃生理盐水；NCU：植物乳杆菌NCU116组，灌胃NCU116植物乳杆菌NCU116（10^9 CFU/mL）生理盐水悬浮液；FCJ：发酵胡萝卜汁组，灌胃植物乳杆菌NCU116发酵胡萝卜汁（含菌量10^9 CFU/mL）；NFCJ：（未发酵）胡萝卜汁组，灌胃未发酵胡萝卜汁。Normal, non-IBD+ saline; IBD, IBD + saline; NCU: IBD + 10^9 CFU/mL L. plantarum NCU116; FCJ: IBD + fermented carrot juice with L. plantarum NCU116; NFCJ: IBD + non-fermented carrot juice.

如图7-3显示，正常组结肠黏膜、隐窝完整，杯状细胞连续分布，部分视野下偶见炎性细胞；结肠炎模型组结肠黏膜糜烂、杯状细胞减少、淋巴细胞浸润。植物乳杆菌NCU116及发酵胡萝卜汁组黏膜上皮细胞有所恢复、杯状细胞增多、炎性细胞减少，其中发酵胡萝卜汁组效果较植物乳杆菌NCU116组更好。未发酵胡萝卜汁组结肠损伤也有部分程度缓解。

7.4 讨 论

本研究利用 TNBS 与乙醇混合溶液来建立结肠炎模型，其发病机制可能是乙醇破坏结肠黏膜结构。TNBS 作为一种半抗原与结肠自身蛋白抗原结合形成完全抗原，从而激活免疫系统，引起免疫应答反应，来达到模型建立的目的。[215, 216] 由图 7-1 和 7-3 可知，本研究成功地复制了结肠炎模型。

有机酸是益生菌发酵的主要终产物，有机酸浓度的提高可以有效降低肠道 pH 和抑制结肠病理损伤。[217] 我们发现，植物乳杆菌 NCU116 发酵胡萝卜汁的短链脂肪酸含量显著高于未发酵胡萝卜汁。并且，丁酸含量在未发酵胡萝卜汁中没有检出，表明植物乳杆菌 NCU116 具有发酵胡萝卜汁中膳食纤维产生特定种类脂肪酸的特性。胡萝卜是日常食用的果蔬之一，其中富含的 β 胡萝卜素具有抗氧化和降低炎症反应的作用。[218, 219] 前期研究发现，发酵后 β 胡萝卜素有少量损失，但大部分被保留。[220]

TNBS 能够导致模型小鼠体重减轻，与其他实验结果类似。虽然实验后期，各给药组都表现出恢复体重的趋势，[221, 222] 但植物乳杆菌 NCU116 及发酵胡萝卜汁组较早表现出这一趋势，提示植物乳杆菌 NCU116 及发酵胡萝卜汁对结肠炎损伤小鼠的体重恢复具有促进作用。

葡萄糖、脂肪酸等在哺乳动物线粒体内代谢可产生自由基，正常机体中自由基处于产生和消除的动态平衡状态。但是，当 TNBS 诱导的结肠炎发生时，机体内会产生复杂的炎性反应，产生的促炎因子会激活活性氧（ROS）过程，导致氧化应激水平升高，主要表现为抗氧化酶活力降低、脂质过氧化物水平升高。[221, 223] 而肠道及黏膜系统极易遭受氧化应激损伤，自由基可诱导肠黏膜中的巯基蛋白变性和不饱和脂肪酸的脂质过氧化，损伤黏膜系统，从而引起结肠炎症。[215] 而植物乳杆菌 NCU116 及发酵胡萝卜汁表现出抑制氧化应激损伤、降低促炎因子和提高促炎因子水平的作用，这些功能对结肠炎损伤修复具有重要意义。

短链脂肪酸主要来自于结肠酵解不被消化的膳食纤维等碳水化合物。短链脂肪酸对宿主机体具有重要的生理功能，如调节肠道菌群平衡、能量供给和维持体液和电解质的平衡等。[224] 本实验中，结肠炎小鼠灌胃植物乳杆菌 NCU116 及发酵胡萝卜汁后，乙酸、丙酸和正丁酸水平均有明显提高。研究显示，短链脂肪酸含量增加能够有提供酸性环境，降低肠道 pH 值，抑制致病菌的生长等多种有益特性。[136, 225]

前期研究发现，植物乳杆菌 NCU116 对便秘小鼠结肠损伤的修复作用可能与短链脂肪酸的产生有关。[185] 文献显示，短链脂肪酸中的丁酸，对结肠上皮细胞具有重要意义，其主要功能有：① 可作为肠上皮细胞的能量来源；② 促进消化道细胞生长；③ 诱导细胞分化；④ 影响基因表达等。[224] 因此，结肠上皮细胞中短链脂肪酸的代谢活动，可能是肠道微生物与宿主机体相互作用最直接的体现。研究显示，结肠内较高的丁酸含量还可能具有抑制大肠癌变的发生。[139, 140] 本实验发现，经植物乳杆菌 NCU116 及发酵胡萝卜汁干预后，小鼠结肠炎症状明显减轻，且两组都表现出修复结肠粘膜损伤的特性。该现象可能与两组中含有较高浓度的短链脂肪酸有关。

7.5　本章小结

通过 TNBS 诱导成功建立了结肠炎小鼠模型，通过植物乳杆菌 NCU116 及发酵胡萝卜汁连续灌胃 5 周后，可知结肠炎小鼠体重较快恢复并有所增长、氧化应激和促炎因子水平降低、结肠粪便短链脂肪酸水平显著升高、结肠黏膜损伤明显降低，其机制可能与短链脂肪酸的含量有关。通过与未发酵胡萝卜汁对比可知，植物乳杆菌 NCU116 及发酵胡萝卜汁对结肠炎的缓解作用与植物乳杆菌 NCU116 的供给密切相关。

第8章 植物乳杆菌NCU116及发酵胡萝卜汁对糖尿病大鼠的血糖改善机制研究

8.1 引 言

Ⅱ型糖尿病是一类以胰岛素抵抗或胰岛素相对缺乏导致高血糖症状的代谢紊乱性疾病。[226] 慢性的高血糖症状可能在多种器官（如眼睛、肾脏、神经、心脏和血管）导致并发症，[227] 从而增加糖尿病患者的死亡率。受基因和环境共同作用，多种因素（如年龄、饮食、缺乏运动和肥胖）都能诱导产生Ⅱ型糖尿病。[228, 229] 过去几十年间，糖尿病患者人数在世界范围内快速增长。当前，已研发出多种治疗糖尿病的药物，如双胍类、噻唑烷二酮类和 α 葡萄糖苷酶类，这些药物能够控制血糖水平，但它们的副作用（肠胃气胀、腹部不适与腹泻等）也不容忽视。[230, 231]

近年来，研究人员逐渐意识到，肠道菌群可能作为一种新颖的干预糖尿病的重要手段。[232, 233] 作为肠道菌群的重要组成部分，益生菌在摄入一定剂量时能够提高机体健康水平。[234, 235] 另外，近期研究发现，益生菌不仅能促进肠道菌群平衡，而且可能具有降血糖功能。[236]

植物乳杆菌NCU116是本实验室分离得到的一株益生菌，它具有降低血清胆固醇等多种活性功能。[101, 185, 237] 作为一种常见的果蔬汁，胡萝卜汁是日常 β 胡萝卜素来源之一。果蔬汁可以以发酵或不发酵的形式饮用。[90] 益生菌发酵蔬菜汁表现出了增强营养和促进人体健康的固有性质，[238] 因而有必要对其可能的机制进行研究。本研究拟采用高脂高糖饲料喂养结合小剂量链脲佐菌素（STZ）建立Ⅱ型糖尿病大鼠模型，研究植物乳杆菌NCU116及其发酵胡萝卜汁对糖尿病大鼠的降血糖、降血脂作用。

8.2 实验部分

8.2.1 实验材料

8.2.1.1 实验动物

Wistar大鼠，雄性，120～150 g，60只，购自常州卡文斯实验动物有限公司，许可证号：SCXK（苏）2011-0003。动物饲料由南昌大学医学院实验动物中心提供。

高脂饲料：66.5%基础饲料、2.5%胆固醇、20%蔗糖、10%猪油、1%胆盐。

饲养环境：温度23 ℃±1 ℃，湿度55%±5%，光暗周期为12 h/12 h光照黑暗交替进行，实验前适应饲养1周，自由饮食饮水。

本实验动物操作获得南昌大学动物实验伦理委员会许可。

8.2.1.2 实验菌种

植物乳杆菌NCU116（南昌大学食品科学与技术国家重点实验室保藏）。

8.2.1.3 试剂耗材

胆固醇（TC）试剂盒（酶法）、甘油三酯（TG）试剂盒（酶法）、低密度脂蛋白胆固醇（LDL-C）试剂盒、高密度脂蛋白胆固醇（HDL-C）试剂盒，北京北化康泰临床试剂有限公司；SOD测定试剂盒、GSH-Px测定试剂盒、MDA测定试剂盒、CAT测定试剂盒、T-AOC测定试剂盒、尿酸试剂盒、肌酐试剂盒、尿素氮试剂盒，南京建成生物工程研究所；苏木精染液、伊红染液，北京中杉金桥生物技术有限公司；cDNA反转录试剂盒，立陶宛Thermo公司；SYBR® Premix Ex Taq™，日本Takara公司；PCR引物，中国Invitrogen公司；链脲佐菌素（STZ），美国Sigma Chemical公司；GLP-1酶联免疫试剂盒、PYY酶联免疫试剂盒，上海心语生物技术公司；胰岛素测定试剂盒、瘦素测定试剂盒、脂联素测定试剂盒，北京华英生物技术研究所。

8.2.2 实验设备

ACCU-CHEK Performa血糖仪，德国罗氏Roche Diagnostics公司；CX31

电子显微镜，日本 Olympus 公司；KD-TS3D 生物组织脱水机、KD2258 生物组织切片机、KD-T 电脑生物组织摊烤片机、D-BM 生物组织包埋机，浙江科迪仪器设备有限公司；7900 HT RT-qPCR 系统，加拿大 Applied Biosystems 公司；R-911 全自动放免计数仪，中国科技大学实业总公司；6890N 气相色谱仪、FID 检测器、HP-INNOWAX 色谱柱，美国 Agilent Technologies 公司；Varioskan Flash 全波长多功能酶标仪，美国 Thermo Scientific 公司。

8.2.3 实验方法

8.2.3.1 糖尿病模型建立与分组

大鼠适应 1 周后，正常组喂以普通饲料，各糖尿病组喂以高脂高糖饲料。各组大鼠自由饮食饮水，每周监测一次体重，8 周后禁食 12 h，实验组大鼠按体重尾静脉注射 30 mg/kg STZ 生理盐水溶液，正常组注射同等剂量生理盐水。1 周后禁食 12 h，尾静脉取血，检测空腹血糖，以连续 3 次空腹血糖值 ≥11.1 mmol/L 并伴有多食、多饮、多尿症状者即为 II 型糖尿病大鼠模型。

将模型建立成功的糖尿病大鼠随机分为：（B）糖尿病模型组（DM），灌胃生理盐水；（C）NCU116 组（NCU），灌胃植物乳杆菌 NCU116（10^9 CFU/mL）生理盐水悬浮液；（D）发酵胡萝卜汁组（FCJ），灌胃含植物乳杆菌 NCU116 发酵胡萝卜汁（含菌量 10^9 CFU/mL）；（E）胡萝卜汁组（NFCJ），灌胃未发酵胡萝卜汁。另设（A）正常组（NDM），灌胃生理盐水。每组 10 只动物，每天按 10 mL/kg 连续给药 5 周。动物自由摄食，自由饮水。

8.2.3.2 发酵胡萝卜制作工艺

参考 7.2.3.2 工艺。

8.2.3.3 空腹血糖的测定

灌胃结束后，大鼠经禁食过夜，尾静脉采血，血糖测定仪和血糖试纸条测定大鼠的空腹血糖。

8.2.3.4 激素水平的检测

大鼠经麻醉，心脏采血，离心得血清。用竞争放免法检测血清胰岛素、胰高血糖素和瘦素含量。取血清，利用 ELISA 法测定 GLP-1 与 PYY 含量。

8.2.3.5 降血脂研究

取上述血清,测定 TC、TG、HDL-C、LDL-C 的含量。

8.2.3.6 氧化应激

参照 2.2.3.7 操作。

8.2.3.7 肾功能

取上述血清,严格按照试剂盒说明书测定尿酸、肌酐和尿素氮含量。

8.2.3.8 粪便短链脂肪酸

参照 2.2.3.4 操作。

8.2.3.9 RT-qPCR 测定糖代谢与脂代谢基因表达

参考 4.2.3.8 操作。测定 LDL receptor,CYP7A1,GLUT-4,PPAR-α,PPAR-γ 基因表达。

8.2.3.10 胰腺和肾脏病理学染色

参考 3.2.3.7 操作。

8.2.3.11 统计学分析

各实验组数据以平均数 ± 标准差($\bar{x}\pm s$)表示,采用 SPSS 17.0 软件进行数据统计分析,Duncan's 多重范围检验。$P < 0.05$ 表示组间具有显著性差异,具有统计学意义。

8.3 结果与分析

8.3.1 血糖水平

如图 8-1 所示,与正常组(4.55 mmol/L)相比,糖尿病模型组(26.72 mmol/L)空腹血糖值显著提高($P < 0.05$)。植物乳杆菌 NCU116 组(19.37 mmol/L)及发酵胡萝汁组(17.19 mmol/L)血糖值较糖尿病模型组显著降低($P < 0.05$)。胡萝卜汁组(23.34 mmol/L)血糖值也有一定程度降低。

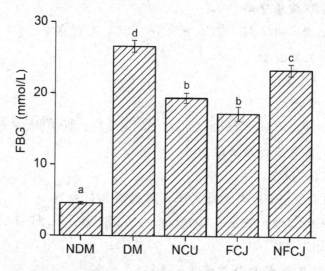

图 8-1 植物乳杆菌 NCU116 及发酵胡萝卜汁对糖尿病大鼠血糖的影响

Figure 8-1 Effect of NCU and FCJ treatment on level of blood glucose in diabetic rats

NDM：正常组，灌胃生理盐水；DM：糖尿病模型组，灌胃生理盐水；NCU：植物乳杆菌 NCU116 组，灌胃 NCU116 植物乳杆菌 NCU116（10^9 CFU/mL）生理盐水悬浮液；FCJ：发酵胡萝卜汁组，灌胃植物乳杆菌 NCU116 发酵胡萝卜汁（含菌量 10^9 CFU/mL）；NFCJ：（未发酵）胡萝卜汁组，灌胃未发酵胡萝卜汁。结果以平均数 ± 标准差表示（$n = 10$），不同上标字母表示组间具有显著性差异（$P < 0.05$）。NDM: non-diabetes mellitus + saline; DM: diabetes mellitus+ saline; NCU: DM + 10^9 CFU/mL *L. plantarum* NCU116; FCJ: DM + fermented carrot juice with *L. plantarum* NCU116; NFCJ: DM + non-fermented carrot juice. Data are expressed as the means ± SEM ($n = 10$). Values with different letters are significantly different ($P < 0.05$).

8.3.2 激素水平

糖尿病模型组血清胰岛素、胰高血糖素和瘦素水平显著高于正常组（$P < 0.05$）。灌胃植物乳杆菌 NCU116 和发酵胡萝卜汁后，3 种激素含量有所降低，其中发酵胡萝卜汁组胰高血糖素和瘦素水平更加接近正常组。胡萝

卜汁组中上述指标与糖尿病模型组没有显著性差异（表 8-1）。糖尿病模型组 GLP-1 和 PYY 水平较正常组显著降低，灌胃植物乳杆菌 NCU116 和发酵胡萝卜汁后，这两种激素水平有所提高（$P < 0.05$）。

表8-1 植物乳杆菌NCU116及发酵胡萝汁对糖尿病大鼠激素水平的影响

Table 8-1 Effect of NCU and FCJ treatment on levels hormones in diabetic rats

Parameters	NDM	DM	NCU	FCJ	NFCJ
Insulin (μIU/ml)	20.64 ± 1.06^a	65.04 ± 4.13^c	39.22 ± 3.49^b	44.54 ± 7.88^b	59.77 ± 2.44^c
Glucagon (pg/ml)	88.26 ± 4.78^a	158.77 ± 10.17^c	130.29 ± 9.10^b	107.50 ± 6.54^a	169.57 ± 4.74^c
Leptin (ng/ml)	4.92 ± 0.71^a	7.33 ± 0.68^b	5.52 ± 0.46^{ab}	4.97 ± 0.30^a	6.48 ± 0.74^{ab}
GLP-1 (pmol/L)	3.42 ± 0.23^c	1.51 ± 0.09^a	2.49 ± 0.11^b	2.81 ± 0.15^b	1.76 ± 0.06^a
PYY (pg/mL)	97.61 ± 3.05^d	41.86 ± 4.21^a	64.02 ± 3.67^b	78.53 ± 2.31^c	44.98 ± 3.37^a

NDM：正常组，灌胃生理盐水；DM：糖尿病模型组，灌胃生理盐水；NCU：植物乳杆菌 NCU116 组，灌胃 NCU116 植物乳杆菌 NCU116（10^9 CFU/mL）生理盐水悬浮液；FCJ：发酵胡萝卜汁组，灌胃植物乳杆菌 NCU116 发酵胡萝卜汁（含菌量 10^9 CFU/mL）；NFCJ:（未发酵）胡萝卜汁组，灌胃未发酵胡萝卜汁。结果以平均数 ± 标准差表示（$n = 10$），不同上标字母表示组间具有显著性差异（$P < 0.05$）。NDM: non-diabetes mellitus + saline; DM: diabetes mellitus+ saline; NCU: DM + 10^9 CFU/mL L. plantarum NCU116; FCJ: DM + fermented carrot juice with L. plantarum NCU116; NFCJ: DM + non-fermented carrot juice. Data are expressed as the means ± SEM ($n = 10$). Values within a row with different letters are significantly different ($P < 0.05$).

8.3.3 血脂水平

表 8-2 显示，糖尿病模型组 TC（8.08 mmol/L）、TG（9.86 mmol/L）和 LDL-C（2.86 mmol/L）较正常组明显提高，HDL-C 水平显著降低，植物乳杆菌 NCU116 组和发酵胡萝卜汁组 TC、TG 和 LDL-C 有所降低，但各糖尿病组 HDL-C 水平差别不明显（0.74 ~ 0.85 mmol/L）。

表8-2 植物乳杆菌NCU116及发酵胡萝卜汁对糖尿病大鼠血脂水平的影响

Table 8-2 Effect of NCU and FCJ treatment on levels of serum lipids in diabetic rats

Parameters	NDM	DM	NCU	FCJ	NFCJ
TC (mmol/L)	1.89 ± 0.10^a	8.08 ± 0.51^c	5.24 ± 0.63^b	6.37 ± 0.65^b	6.75 ± 0.36^{bc}
TG (mmol/L)	1.10 ± 0.10^a	9.86 ± 0.56^c	7.31 ± 0.81^b	7.47 ± 0.37^b	8.60 ± 0.56^{bc}
HDL-C (mmol/L)	1.00 ± 0.04^b	0.74 ± 0.03^a	0.85 ± 0.07^a	0.81 ± 0.05^a	0.80 ± 0.02^a
LDL-C (mmol/L)	0.39 ± 0.09^a	2.86 ± 0.32^d	1.07 ± 0.20^{ab}	1.54 ± 0.19^{bc}	2.04 ± 0.34^c

NDM：正常组，灌胃生理盐水；DM：糖尿病模型组，灌胃生理盐水；NCU：植物乳杆菌NCU116组，灌胃NCU116植物乳杆菌NCU116（10^9 CFU/mL）生理盐水悬浮液；FCJ：发酵胡萝卜汁组，灌胃植物乳杆菌NCU116发酵胡萝卜汁（含菌量10^9 CFU/mL）；NFCJ：（未发酵）胡萝卜汁组，灌胃未发酵胡萝卜汁。结果以平均数±标准差表示（$n = 10$），不同上标字母表示组间具有显著性差异（$P < 0.05$）。NDM: non-diabetes mellitus + saline; DM: diabetes mellitus+ saline, NCU: DM + 10^9 CFU/mL *L. plantarum* NCU116, FCJ: DM + fermented carrot juice with *L. plantarum* NCU116; NFCJ: DM + non-fermented carrot juice. Data are expressed as the means ± SEM ($n = 10$). Values within a row with different letters are significantly different ($P < 0.05$).

8.3.4 氧化应激

糖尿病模型组SOD（156.94 U/mL）、GSH-Px（289.28 U/mL）、CAT（98.65 U/mL）活力和T-AOC（6.67 U/mL）水平显著低于正常组（$P < 0.05$，表8-3）。发酵胡萝卜汁组中这4项指标（分别为200.35、361.72、150.32和13.35 U/mL）有所提高。植物乳杆菌NCU116组（3.73 nmol/mL）和发酵胡萝卜汁组（3.70 nmol/mL）MDA含量低于糖尿病模型组（5.36 nmol/mL）和胡萝卜汁组（5.74 nmol/mL）。

表8-3 植物乳杆菌NCU116及发酵胡萝卜汁对糖尿病大鼠氧化应激水平的影响

Table 8-3 Effect of NCU and FCJ treatment on oxidative stress in diabetic rats

Parameters	NDM	DM	NCU	FCJ	NFCJ
SOD (U/mL)	219.53 ± 4.76^c	156.94 ± 6.03^a	191.29 ± 6.20^b	200.35 ± 8.95^b	166.43 ± 3.61^a
GSH-Px (U/mL)	402.73 ± 16.18^b	289.28 ± 34.29^a	353.65 ± 4.22^b	361.72 ± 9.71^b	293.52 ± 10.86^a
MDA (nmol/mL)	3.59 ± 0.19^a	5.36 ± 0.41^{ab}	3.73 ± 0.14^a	3.70 ± 0.17^a	5.74 ± 1.21^b
T-AOC (U/mL)	14.77 ± 0.80^c	6.67 ± 0.56^a	11.62 ± 0.63^b	13.35 ± 0.66^{bc}	8.23 ± 0.83^a
CAT (U/mL)	198.59 ± 12.36^c	98.65 ± 7.13^a	120.18 ± 9.67^{ab}	150.32 ± 18.96^b	107.48 ± 7.00^a

NDM：正常组，灌胃生理盐水；DM：糖尿病模型组，灌胃生理盐水；NCU：植物乳杆菌NCU116组，灌胃NCU116植物乳杆菌NCU116（10^9 CFU/mL）生理盐水悬浮液；FCJ：发酵胡萝卜汁组，灌胃植物乳杆菌NCU116发酵胡萝卜汁（含菌量10^9 CFU/mL）；NFCJ：（未发酵）胡萝卜汁组，灌胃未发酵胡萝卜汁。结果以平均数 ± 标准差表示（$n = 10$），不同上标字母表示组间具有显著性差异（$P < 0.05$）。NDM: non-diabetes mellitus + saline; DM: diabetes mellitus+ saline; NCU: DM + 10^9 CFU/mL *L. plantarum* NCU116; FCJ: DM + fermented carrot juice with *L. plantarum* NCU116; NFCJ: DM + non-fermented carrot juice. Data are expressed as the means ± SEM ($n = 10$). Values within a row with different letters are significantly different ($P < 0.05$).

8.3.5 肾功能

高脂饲料喂养结合STZ建立II型糖尿病大鼠模型会升高血清尿素氮、尿酸和肌酐水平。灌胃植物乳杆菌NCU116和发酵胡萝卜汁后，尿素氮和肌酐水平明显降低（$P < 0.05$），尿酸水平也有所降低，但结果不具有显著性差异（$P > 0.05$），见表8-4。

表8-4 植物乳杆菌NCU116及发酵胡萝汁对糖尿病大鼠肾功能的影响
Table 8-4 Effect of NCU and FCJ treatment on kidney function in diabetic rats

Parameters	NDM	DM	NCU	FCJ	NFCJ
Urea nitrogen (mmol/L)	2.79 ± 0.36^a	13.30 ± 1.87^c	9.25 ± 0.24^b	6.84 ± 0.67^b	7.82 ± 0.86^b
Creatinine (mmol/L)	0.16 ± 0.01^a	0.98 ± 0.05^c	0.58 ± 0.06^b	0.44 ± 0.08^b	0.42 ± 0.04^b
Uric acid (mmol/L)	0.26 ± 0.04^a	0.56 ± 0.08^b	0.48 ± 0.10^{ab}	0.43 ± 0.12^{ab}	0.51 ± 0.08^{ab}

NDM：正常组，灌胃生理盐水；DM：糖尿病模型组，灌胃生理盐水；NCU：植物乳杆菌NCU116组，灌胃NCU116植物乳杆菌NCU116（10^9 CFU/mL）生理盐水悬浮液；FCJ：发酵胡萝卜汁组，灌胃植物乳杆菌NCU116发酵胡萝卜汁（含菌量10^9 CFU/mL）；NFCJ：（未发酵）胡萝卜汁组，灌胃未发酵胡萝卜汁。结果以平均数 ± 标准差表示（$n = 10$），不同上标字母表示组间具有显著性差异（$P < 0.05$）。NDM: non–diabetes mellitus + saline; DM: diabetes mellitus+ saline; NCU: DM + 10^9 CFU/mL *L. plantarum* NCU116; FCJ: DM + fermented carrot juice with *L. plantarum* NCU116; NFCJ: DM + non–fermented carrot juice. Data are expressed as the means ± SEM ($n = 10$). Values within a row with different letters are significantly different ($P < 0.05$).

8.3.6 粪便短链脂肪酸含量

糖尿病模型组含有乙酸（120.54 μmol/g）、丙酸（49.15 μmol/g）、丁酸（16.44 μmol/g），较正常组低（$P < 0.05$）。植物乳杆菌NCU116组、发酵胡萝卜汁组和胡萝卜汁组短链脂肪酸含量有所提高，其中植物乳杆菌NCU116组和发酵胡萝卜汁组较糖尿病模型组有显著性差异（$P < 0.05$）。植物乳杆菌NCU116组和发酵胡萝卜汁组总短链脂肪酸水平接近正常组（图8-2）。

NDM：正常组，灌胃生理盐水；DM：糖尿病模型组，灌胃生理盐水；NCU：植物乳杆菌NCU116组，灌胃NCU116植物乳杆菌NCU116（10^9 CFU/mL）生理盐水悬浮液；FCJ：发酵胡萝卜汁组，灌胃植物乳杆菌NCU116发酵胡萝卜汁（含菌量10^9 CFU/mL）；NFCJ：（未发酵）胡萝卜汁组，灌胃未发酵胡萝卜汁。结果以平均数 ± 标准差表示（$n = 10$），不同上标字母表示组间具有显著性差异（$P < 0.05$）。NDM: non-diabetes mellitus + saline; DM: diabetes mellitus+ saline; NCU: DM + 10^9 CFU/mL *L. plantarum* NCU116; FCJ: DM + fermented carrot juice with *L. plantarum* NCU116; NFCJ: DM + non-fermented carrot juice. Data are expressed as the means ± SEM ($n = 10$). Values with different letters are significantly different ($P < 0.05$).

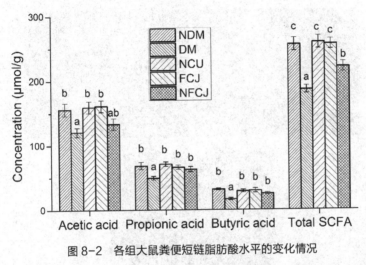

图 8-2　各组大鼠粪便短链脂肪酸水平的变化情况

Figure 8-2 Changes in short chain fatty acids (SCFA) in feces of five groups of rats.

8.3.7 胰腺与肾脏病理学

正常组胰腺组织腺泡细胞规则地排列在胰岛细胞周围、胰岛形态正常；糖尿病模型组胰岛 β 细胞显著减少，部分胰岛退化萎缩；植物乳杆菌 NCU116 组和发酵胡萝卜汁组胰岛细胞损伤减轻、空泡化减少、胰岛数量增加（图 8-3 A）。

与之相似的，与正常组比较，糖尿病大鼠肾脏的系膜增宽、肾小管间质纤维化、肾小球上皮细胞空泡样变性、肾小球硬化萎缩等症状。植物乳杆菌 NCU116 组和发酵胡萝卜汁组肾小球间质纤维化程度降低、上皮细胞空泡症状减轻、肾小球萎缩症状有所缓解（图 8-3 B）。

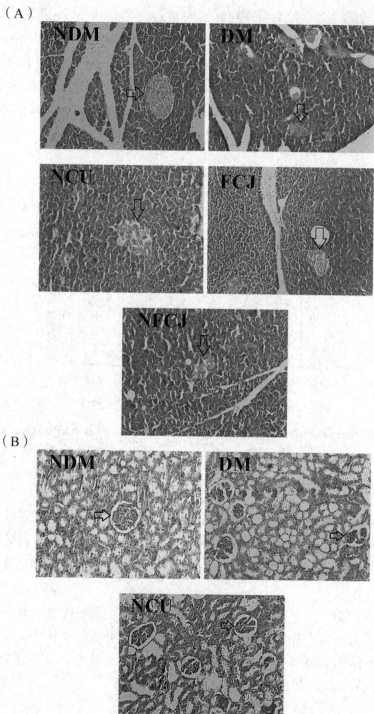

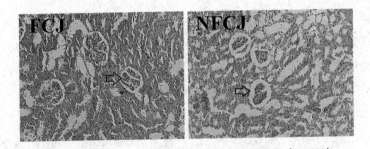

图 8-3 胰腺（A）和肾脏（B）组织病理学变化（100X）

Figure 8-3 Histological findings in the pancreas (A) and kidney (B)

NDM：正常组，灌胃生理盐水；DM：糖尿病模型组，灌胃生理盐水；NCU：植物乳杆菌NCU116组，灌胃NCU116植物乳杆菌NCU116（10^9 CFU/mL）生理盐水悬浮液；FCJ：发酵胡萝卜汁组，灌胃植物乳杆菌NCU116发酵胡萝卜汁（含菌量 10^9 CFU/mL）；NFCJ：（未发酵）胡萝卜汁组，灌胃未发酵胡萝卜汁。NDM: non-diabetes mellitus + saline; DM: diabetes mellitus+ saline; NCU: DM + 10^9 CFU/mL *L. plantarum* NCU116; FCJ: DM + fermented carrot juice with *L. plantarum* NCU116，NFCJ: DM + non-fermented carrot juice.

8.3.8 脂代谢与糖代谢基因表达

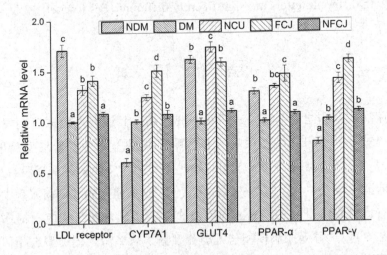

图 8-4 植物乳杆菌NCU116及发酵胡萝汁对糖尿病大鼠脂代谢与糖代谢的影响

Figure 8-4 Effect of NCU and FCJ treatment on mRNA expression of lipid and glucose metabolism.

与正常组比较，糖尿病大鼠肝脏 LDL 受体和 PPAR-α 以及骨骼肌 GLUT4 基因表达降低（$P < 0.05$），灌胃植物乳杆菌 NCU116 和发酵胡萝卜汁 5 周后，上述基因表达水平显著提高。同时，两个给药组 CYP7A1 和 PPAR-γ 基因表达水平显著高于糖尿病模型组（图 8-4）。

NDM：正常组，灌胃生理盐水；DM：糖尿病模型组，灌胃生理盐水；NCU：植物乳杆菌 NCU116 组，灌胃 NCU116 植物乳杆菌 NCU116（10^9 CFU/mL）生理盐水悬浮液；FCJ：发酵胡萝卜汁组，灌胃植物乳杆菌 NCU116 发酵胡萝卜汁（含菌量 10^9 CFU/mL）；NFCJ：（未发酵）胡萝卜汁组，灌胃未发酵胡萝卜汁。结果以平均数 ± 标准差表示（$n = 10$），不同上标字母表示组间具有显著性差异（$P < 0.05$）。NDM: non-diabetes mellitus + saline; DM: diabetes mellitus+ saline; NCU: DM + 10^9 CFU/mL *L. plantarum* NCU116; FCJ: DM + fermented carrot juice with *L. plantarum* NCU116, NFCJ: DM + non-fermented carrot juice. LDL receptor: Low-density lipoprotein (LDL) receptor; CYP7A1: Cholesterol 7α-hydroxylase; GLUT-4: Glucose transporter-4; PPAR-α: Peroxisome proliferator-activated receptor-α; PPAR-γ: Peroxisome proliferator-activated receptor-γ. The graph represents the mRNA levels relative to β-actin. Data are expressed as the means ± SEM ($n = 10$). Values with different letters are significantly different ($P < 0.05$).

8.4 讨 论

糖尿病主要包含Ⅰ型糖尿病和Ⅱ型糖尿病，其中Ⅱ型糖尿病的发病率（90%~95%）远高于Ⅰ型糖尿病。Ⅱ型糖尿病是非胰岛素依赖型糖尿病，以高血糖、胰岛素抵抗和胰岛素相对缺乏（非绝对缺乏）为主要特征。[227] 高脂高糖饲料喂养结合小剂量链脲佐菌素（STZ）建立的Ⅱ型糖尿病大鼠模型，具有血糖水平升高、胰岛 β 细胞损伤、血脂异常和氧化应激损伤等特点（图 8-5）。这种糖尿病模型与人类Ⅱ型糖尿病患者代谢类型有相似的特点，因而该实验性糖尿病动物模型研究中具有广泛的应用价值。[239]

本研究中，糖尿病大鼠表现出高血糖和胰岛素抵抗症状（图 8-1，表 8-1），结果显示植物乳杆菌 NCU116 及其发酵胡萝卜汁能够在一定程度缓解高血糖和胰岛素抵抗。

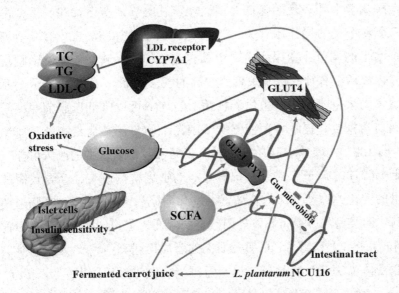

图 8-5 植物乳杆菌 NCU116 及发酵胡萝汁对高脂高糖饲料喂养结合小剂量链脲佐菌素诱导建立 II 型糖尿病大鼠模型的可能作用机制

Figure 8-5 Possible mechanism involved in anti-diabetic effects of L. plantarum NCU116 and its fermented carrot juice on a high fat fed and low-dose streptozotocin -induced type 2 diabetic rat model.

前期研究发现,植物乳杆菌 NCU116 发酵胡萝卜汁的短链脂肪酸含量显著高于未发酵胡萝卜汁。本部分将重点探讨短链脂肪酸对糖尿病大鼠的影响。高脂饮食可以降低短链脂肪酸的浓度,但结肠内菌群发酵能够补偿这种失衡。[240, 241] 植物乳杆菌 NCU116 具有体外产短链脂肪酸的性质,本实验对其在糖尿病大鼠体内产短链脂肪酸的性能进行了研究。结果显示,糖尿病大鼠粪便短链脂肪酸含量明显降低。灌胃植物乳杆菌 NCU116 及其发酵胡萝卜汁后,粪便短链脂肪酸含量显著升高。短链脂肪酸是乳酸杆菌和双歧杆菌的关键代谢产物,它们具有多种生物学性质,可在多个通路上抑制糖尿病症状。[242] 其中,短链脂肪酸可以降低血糖、胰岛素抵抗、炎症并能提升 PYY 和 GLP-1 的分泌。[240]

丙酸可以降低肝脏细胞内葡萄糖的产生。[243] 本研究中,植物乳杆菌

NCU116 和胡萝卜汁都能升高粪便丙酸含量，但发酵胡萝卜汁表现的更为明显。发酵胡萝卜汁表现的降低血糖性能可能与 β 胡萝卜素和肠道菌群产生的短链脂肪酸的协同作用有关。[89]另外，激素（如胰岛素、胰高血糖素和瘦素）水平的调节在糖尿病模型中具有重要作用。胰岛素能够影响瘦素分泌，并且代谢综合征大鼠常表现为较高的瘦素症状。[160]前期研究显示，在高脂饮食的大鼠模型中，植物乳杆菌 NCU116 可以降低口服葡萄糖耐受水平和胰岛素抵抗，提示该菌能够调节胰岛细胞分泌胰岛素和胰高血糖素水平。[185]因此，本实验中胰岛素、胰高血糖素和瘦素水平的改善可能与植物乳杆菌 NCU116 调节胰岛素抵抗有关。肠道菌群的改变，会导致短链脂肪酸含量的变化，这些变化与胰岛素抵抗和糖尿病密切相关。[240]肠道菌群组成及其代谢类型的改变可影响外周器官的代谢进程，如肠道激素的分泌（如 PYY 和 GLP-1）和脂肪、肝脏及肌肉的脂质代谢类型。[244]研究证实，G 蛋白偶联受体（GPR）41 缺乏小鼠经短链脂肪酸刺激后能促进肠道激素 PYY 的释放。该肽类激素表现出降低肠道推进时间的性能，表明它可以促进营养物质（如葡萄糖）的吸收。[245]丙酸和丁酸可降低炎症水平并诱导激素释放。在人与动物体内，短链脂肪酸与 GLP-1 的分泌关系密切。GLP-1 是一种由肠道分泌并参与调控血糖平衡的激素，它能降低血糖浓度与保持胰岛 β 细胞功能，[240]其抗糖尿病的机制可能是该激素具有刺激葡萄糖依赖性胰岛素释放和促进血糖平衡的作用。[232, 246]

血脂异常是糖尿病的常见特征。[231]在 STZ 诱导的糖尿病大鼠中，主要表现为 TC、TG 和 LDL-C 水平升高与 HDL-C 水平降低，灌胃经 *L. rhamnosus* CRL981 发酵的豆浆后，血脂水平显著改善。[238]高水平的血清 TC 和 LDL-C 是诱发心血管病的潜在因素。降低血清 TC 和 LDL-C 水平，可以降低患动脉粥样硬化的风险。前期研究表明，植物乳杆菌 NCU116 可以显著降低高脂饮食大鼠的 TC、TG 和 LDL-C 水平，并能改善 HDL-C 水平。[185]粪便移植术的研究证实，肠道菌群产生的丁酸在调节血脂与血糖平衡中具有重要作用。[240]本实验中，发酵胡萝卜汁组表现出最好的调节 TC、TG 和 LDL-C 水平的能力，可能与其肠道内丁酸含量较高有关。

氧化应激在糖尿病中具有重要的研究价值。高血糖和高血脂症状具有损伤抗氧化系统的风险。[247]糖尿病与氧化应激的关系，从本质上来讲是自由基与抗氧化系统失衡引发的。[248]氧化应激能够损伤细胞膜，这也可能是诱导胰岛和肾小球损伤的主要原因。STZ 诱导的糖尿病因过氧化氢和超氧自由基等活性氧（ROS）的积累，导致该模型具有低 SOD、GSH-Px、CAT 和 T-AOC 的特点。[249]MDA 是脂质过氧化损伤的终产物和标志物，具有 DNA 和蛋白质毒性。本研究发现，植物乳杆菌 NCU116 及其发酵胡萝卜汁显著改善氧化应激水平，可能是植物乳杆菌 NCU116 和 β 胡萝卜素共同的作用结果。

糖尿病的高血糖、高血脂与自由基的生成密切相关。氧化性损伤导致胰腺和肾脏的机能障碍，植物乳杆菌 NCU116 和发酵胡萝卜汁灌胃后，胰腺 β 细胞坏死和空泡化现象减轻，肾脏微观结构显著改善。[250] 目前为止，少有研究表明，发酵胡萝卜汁可以缓解病理损伤，但有研究探讨了 β 胡萝卜素的补充与糖尿病发病率的关系。[251]

长时间的高血糖症状可诱发氧化应激、组织病理损伤，从而产生肾脏等组织的功能紊乱。肾脏的机能障碍引发血液内代谢毒性物质（如尿酸、肌酐、尿素氮）堆积。[252] 从形态学上来讲，植物乳杆菌 NCU116 和发酵胡萝卜汁具有修复肾脏病理损伤的作用，肾功能的改善可能与肾脏的病理修复有关。

肠道微生物不仅能影响短链脂肪酸的产生，而且能影响宿主脂肪酸吸收、氧化和储存的基因表达。[245] 前期研究表明，植物乳杆菌 NCU116 可以上调高脂饮食大鼠 LDL 受体和 CYP7A1 的基因表达，来降低血清 TC 和 LDL-C 含量。[185] 本研究发现，植物乳杆菌 NCU116 和发酵胡萝卜汁能显著提升 LDL 受体和 CYP7A1 表达，这可能与血清 LDL-C 的降低有关。GLUT-4 是存在于脂肪和骨骼肌组织的葡萄糖转运蛋白。[253] 骨骼肌 GLUT-4 在植物乳杆菌 NCU116 和发酵胡萝卜汁的作用下，表达水平显著提高。80% 以上受胰岛素支配的葡萄糖在骨骼肌内消耗，骨骼肌 GLUT-4 表达水平的提高可以增加糖尿病患者的葡萄糖摄取率。[254]PPAR 是一类重要的调节炎性反应和葡萄糖平衡的基因。PPAR-α 可能具有提高胰岛素敏感性的性能；PPAR-γ 是药物治疗糖尿病的靶标基因。[255, 256] 植物乳杆菌 NCU116 和发酵胡萝卜汁摄入后，通过调控这些基因的表达，可以促进葡萄糖稳态和提升胰岛素敏感性。上述基因的表达调控是植物乳杆菌 NCU116 和发酵胡萝卜汁调节高脂饮食结合小剂量 STZ 诱导 II 型糖尿病的可能机制。

8.5 本章小结

（1）植物乳杆菌 NCU116 及发酵胡萝卜汁具有血糖、血脂、激素与氧化应激的调节能力。

（2）植物乳杆菌 NCU116 及发酵胡萝卜汁具有提高粪便短链脂肪酸水平，修复胰腺组织损伤以及调节糖代谢与脂代谢基因表达的性能。

（3）初步证实植物乳杆菌 NCU116 及发酵胡萝卜汁具有一定的缓解糖尿病症状的功能。

第9章 植物乳杆菌NCU116及发酵胡萝卜汁对糖尿病大鼠的血清代谢组学初探

9.1 引 言

代谢组学定义为"通过充分定量的代谢物来综合评估涉及生物样本的化学过程,并寻找代谢物和生理变化的关系",它是系统生物学继基因组学、转录组学、蛋白质组学之后新兴的一门组学技术。[95, 257]代谢组学在疾病诊断、药物研发、通路识别、营养学、微生物学研究等诸多领域得到广泛应用。[96, 97]小分子代谢物质(如脂肪酸、氨基酸、多肽、有机酸、核酸和维生素)为系统生物提供基本信息来帮助理解疾病表型。[258]当前,有多种仪器可用来进行代谢组学分析。在多种分析平台当中,超高压液相色谱-四级飞行时间质谱(UPLC-Q-TOF/MS)因其高效的性能,越来越受到科研工作者的青睐。[257, 259]

糖尿病是一类以胰岛素抵抗和相对缺乏导致的高血糖为主要症状的代谢综合征。[227]近年来,代谢组学技术常被用来研究包含糖尿病、肥胖和心血管病在内的代谢综合征。[238, 260]研究发现,益生菌不仅能够促进肠道健康,还能有效缓解代谢综合征症状。[261, 262]

植物乳杆菌NCU116是一株体内外活性较强的益生菌。[101, 185, 237]胡萝卜汁是日常β胡萝卜素的重要来源。[263]因此,本实验拟研究植物乳杆菌NCU116及其发酵胡萝卜汁的促营养与健康性能。采用UPLC-Q-TOF/MS来分析植物乳杆菌NCU116和发酵胡萝卜汁在Ⅱ型糖尿病大鼠模型中的代谢物与通路,并利用非配对t检验分析、建立主成分分析和偏最小二乘法分析模型来研究各组代谢物的差异和分布规律。

9.2 实验部分

9.2.1 实验材料

9.2.1.1 实验动物

Wistar 大鼠，雄性，120～150 g，60 只，购自常州卡文斯实验动物有限公司，许可证号：SCXK（苏）2011-0003。动物饲料由南昌大学医学院实验动物中心提供。

高脂饲料：66.5% 基础饲料、2.5% 胆固醇、20% 蔗糖、10% 猪油、1% 胆盐。

饲养环境：温度 23 ℃ ±1 ℃，湿度 55% ±5%，光暗周期为 12 h/12 h 光照黑暗交替进行，实验前适应饲养 1 周，自由饮食饮水。

本实验动物操作获得南昌大学动物实验伦理委员会许可。

9.2.1.2 实验菌种

植物乳杆菌 NCU116（南昌大学食品科学与技术国家重点实验室保藏）。

9.2.1.3 试剂耗材

链脲佐菌素（STZ）、甲酸（HPLC grade），美国 Sigma 公司；乙腈（HPLC grade），德国 Merck 公司；纯净水，广州屈臣氏公司。

9.2.2 实验设备

1290 系列 UPLC、6538 系列 Q-TOF/MS、ESI 电喷雾离子源、Eclipse Plus C_{18} 色谱柱（3.5 μm, 2.1 mm i.d.×150 mm）、Mass Hunter Qualitative Analysis 软件、Mass Profiler Professional（MPP, B.02.00）软件，美国 Agilent Technologies 公司；代谢笼，苏州苏杭科技公司；冷冻离心机，美国 Sigma 公司；ACCU-CHEK Performa 血糖仪，德国 Roche Diagnostics 公司。

9.2.3 实验方法

9.2.3.1 糖尿病模型建立与分组

大鼠适应 1 周后，正常对照组喂以普通饲料，实验组喂以高脂高糖饲料。各组大鼠自由饮食饮水,每周监测一次体重,8 周后禁食 12 h,实验组大鼠按体重尾静脉注射 30 mg/kg STZ 生理盐水溶液,正常组注射同等剂量生理盐水。1 周后禁食 12 h,尾静脉取血,检测空腹血糖,以连续 3 次空腹血糖值 ≥ 11.1 mmol/L 并伴有多食、多饮、多尿症状者即为 II 型糖尿病大鼠模型。

将 40 只成模糖尿病大鼠随机分为:(B)糖尿病模型组(DM),灌胃生理盐水;(C)NCU116 组(NCU),灌胃植物乳杆菌 NCU116(10^9 CFU/mL)生理盐水悬浮液;(D)发酵胡萝卜汁组(FCJ),灌胃含植物乳杆菌 NCU116 发酵胡萝卜汁(含菌量 10^9 CFU/mL);(E)胡萝卜汁组(NFCJ),灌胃未发酵胡萝卜汁。另设(A)正常组,灌胃生理盐水。每组 10 只动物,每天按 10 mL/kg 连续给药 5 周。动物自由摄食,自由饮水。

9.2.3.2 发酵胡萝卜汁的制备

参考 7.2.3.2 工艺操作。

9.2.3.3 样品收集与前处理

灌胃结束后,大鼠经麻醉,心脏采血,离心得血清。取 200 μL 血清加入 800 μL 乙腈-去离子水(4:1,v/v)中,涡旋混匀,离心,上清液过膜后,置于样瓶中。

9.2.3.4 UPLC-Q-TOF/MS 分析

参考 6.2.3.3 操作。

9.2.3.5 数据处理与统计分析

参考 6.2.3.4 操作。

9.3 结果与分析

9.3.1 血清代谢物图谱分析

UPLC-Q-TOF/MS 是一种强有力的代谢组学分析工具。[264] 本研究中,利用 UPLC-Q-TOF/MS 对血清样品进行正(负)模式离子扫描来监测各组代谢物的差异。如图 9-1 所示,正离子扫描模式(A)及负离子扫描模式(B)下的总离子流色谱图,显示正负离子模式都具有较好的分离度和离子丰度。另外，研究采用正离子（m/z 值为 121.050 8 和 922.009 7）和负离子（m/z 值为 112.398 55 和 1034.988 1）的参比离子来确保正负离子扫描的质量准确性和重现性。

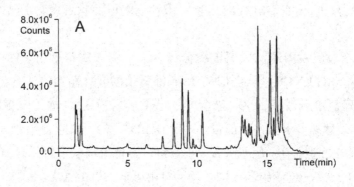

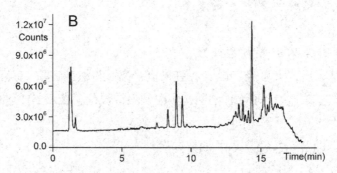

图 9-1　Ⅱ型糖尿病大鼠血清样品总离子流图，（A）正离子模式，（B）负离子模式
Figure 9-1 Positive (A) and negative (B) ion chromatograms of the serum sample from a type 2 diabetic rat.

9.3.2 代谢物多元变量分析

在 UPLC-Q-TOF/MS 分析的基础上，为了更好地体现 5 组给药成分的代谢差异，首先对实验结果进行 PCA 分析。[204] 图 9-2 显示为 PCA 分析正离子（A）和负离子（B）的 3D 得分图，图中每个点代表一个样本，每个样本小分子代谢物的成分和浓度决定该样本在图中的位置，成分和浓度相近的样本，在得分图上的位置也比较靠近。结果显示，与正常组相比，糖尿病模型组分布位置较远，提示糖尿病模型组大鼠样本化合物组分和浓度与正常组有较大差异，其体内生物化学过程有了明显改变。与糖尿病模型组相比，植物乳杆菌 NCU116 和发酵胡萝卜汁组有向正常组靠近的趋势，说明植物乳杆菌 NCU116 和发酵胡萝卜汁可以调节糖尿病大鼠血液中小分子化合物的组分与含量。另外，正、负离子模式下 PCA 8 种主成分经 MPP 软件分析分别为 68.7% 和 71.38%。

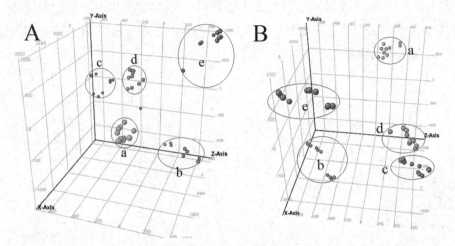

图 9-2　各组 3D PCA 得分图：（A）正离子模式，（B）负离子模式

Figure 9-2 Score plots in positive (A) and negative (B) ion mode from 3D PCA model classifying five groups

　　a：正常组，灌胃生理盐水；b：糖尿病模型组，灌胃生理盐水；c：植物乳杆菌 NCU116 组，灌胃 NCU116 植物乳杆菌 NCU116（10^9 CFU/mL）生理盐水悬浮液；d：发酵胡萝卜汁组，灌胃植物乳杆菌 NCU116 发酵胡萝卜汁（含菌量 10^9 CFU/mL）；e：（未发酵）胡萝卜汁组，灌胃未发酵胡萝卜汁。$n = 10$。a: NDM, non-diabetes mellitus; b: DM, diabetes mellitus; c: NCU, DM + *L. plantarum* NCU116; d: FCJ, DM + fermented carrot juice with *L. plantarum* NCU116; e: NFCJ, DM + non-fermented carrot juice. $n = 10$ in each group.

　　a：正常组，灌胃生理盐水；b：糖尿病模型组，灌胃生理盐水；c：植物乳杆菌 NCU116 组，灌胃 NCU116 植物乳杆菌 NCU116（10^9 CFU/mL）生理盐水悬浮液；d：发酵胡萝卜汁组，灌胃植物乳杆菌 NCU116 发酵胡萝卜汁（含菌量 10^9 CFU/mL）；e：（未发酵）胡萝卜汁组，灌胃未发酵胡萝卜汁。$n = 10$。a: NDM, non-diabetes mellitus; b: DM, diabetes mellitus; c: NCU, DM + *L. plantarum* NCU116; d: FCJ, DM + fermented carrot juice with *L. plantarum*

NCU116; e: NFCJ, DM + non-fermented carrot juice. n = 10 in each group.

与 PCA 分析相比，PLS-DA 是一种更清晰地反映组间差异的一种监督多变量的统计学方法。[205]PLS-DA 分析结果显示，植物乳杆菌 NCU116、发酵胡萝卜汁和胡萝卜汁组样本间差异更加明显（图 9-3）。因此，植物乳杆菌 NCU116 和发酵胡萝卜汁改变了糖尿病大鼠的血清代谢类型。发酵胡萝卜汁组样本更为靠近正常组，表明发酵胡萝卜汁组在改善糖尿病血清代谢方面表现最为明显。

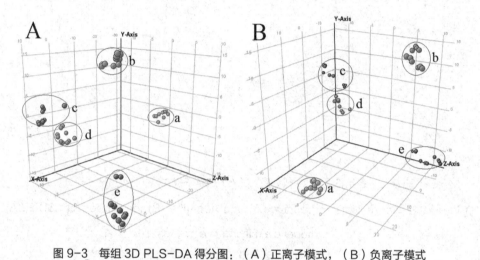

图 9-3　每组 3D PLS-DA 得分图：（A）正离子模式，（B）负离子模式

Figure 9-3 Score plots in positive (A) and negative (B) ion mode from 3D PLS-DA model classifying five groups

交叉验证用来评估 PLS-DA 模型的准确性。经验证矩阵模型分析可知，正离子扫描模式下准确率为 98%（植物乳杆菌 NCU116 组为 90%，其他组为 100%）；负离子扫描模式下准确率为 100%（各组均为 100%）。该验证矩阵模型数值在接近 100% 的情况下，表示该数学模型建立成功，可以相当准确地预测结果。[205]此外，该数据还显示各组具有很好的分离度，表示糖尿病大鼠经植物乳杆菌 NCU116、发酵胡萝卜汁和胡萝卜汁组灌胃后，血清小分子代谢物成分发生了较大变化。灌胃 5 周后，与糖尿病组比较植物乳杆菌 NCU116 和发酵胡萝卜汁组更为靠近正常组，提示植物乳杆菌 NCU116 和发酵胡萝卜汁在一定程度上影响了糖尿病大鼠的代谢进程。本结果与前期研究（植物乳杆菌 NCU116 和发酵胡萝卜汁对糖尿病大鼠具有降血糖和

降血脂作用）一致。[265]

9.3.3 潜在生物标记物分析

数据库的系统搜索分析对潜在生物标记物的鉴定具有重要意义。[208] 通过非配对 t 检验来分析不同的生物标记物。利用筛选差异化合物流程，通过 Agilent METLIN Personal Metabolite Database 对标记物 ID 进行识别，匹配标记物 CAS 号，并利用 ID Browser 识别功能确定可能的生物标记物（表 9-1）。

表9-1 血清潜在生物标记物与其变化趋势

Table 9-1 Metabolites selected as biomarkers characterized in serum profile and their change trends

Ion model	Compound Name	DM*	NCU#	FCJ#	NFCJ#
Positive	Adenosine	↑	↓	↓	↓
	Proline	↑	↓	↓	↓
	5-Hydroxy indole acetaldehyde	↓	--	--	↑
	Glycocholic Acid	↑	↓	↓	↓
	Taurochenodeoxycholic acid	↑	↓	↓	↓
	Sphingosine	↑	↑	↑	↑
Negative	Theophylline	↑	↓	↓	↓
	Taurocholic acid	↑	↓	↓	↓

NDM：正常组，灌胃生理盐水；DM：糖尿病模型组，灌胃生理盐水；NCU：植物乳杆菌 NCU116 组，灌胃 NCU116 植物乳杆菌 NCU116（10^9 CFU/mL）生理盐水悬浮液；FCJ：发酵胡萝卜汁组，灌胃植物乳杆菌 NCU116 发酵胡萝卜汁（含菌量 10^9 CFU/mL）；NFCJ：（未发酵）胡萝卜汁组，灌胃未发酵胡萝卜汁。* 表示与正常组相比的变化趋势，# 表示与模型组相比的变化趋势。↓ 表示代谢物含量下调，↑ 表示代谢物含量上调。$n = 10$。NDM, non-diabetes mellitus; DM, diabetes mellitus; NCU, DM + *L. plantarum* NCU116; FCJ, DM + fermented carrot juice with *L. plantarum*

NCU116; NFCJ, DM + non-fermented carrot juice. * Change trend compared with NDM group. # Change trend compared with DM group. The levels of potential biomarkers were labeled with (↓) down-regulated and (↑) up-regulated. n=10 in each group.

通过 MPP 软件分析发现，经植物乳杆菌 NCU116 和发酵胡萝卜汁干预可明显改变一些代谢过程。利用 UPLC-Q-TOF/MS/MS 的二级离子碎片来鉴别潜在的生物标记物。[266,267] 通过碰撞电压为 10~40 eV 得到的 MS/MS 图谱碎片离子的信息导出为 CEF 格式文件，通过 MSC 软件和 Chemspider 数据库分析，最终鉴定本研究中的生物标记物。分析发现，正离子特征代谢标志物主要有腺苷（adenosine）、5-羟吲哚乙醛（5-hydroxy indole acetaldehyde）、脯氨酸（proline）、甘氨胆酸（glycocholic acid）、牛磺鹅去氧胆酸（taurochenodeoxycholic acid）、鞘氨醇（sphingosine）；负离子特征代谢标记物主要有茶碱（theophylline）和牛磺胆酸（taurocholic acid）。

9.4 讨 论

表 9-2 和表 9-3 显示，利用 Pathway Analysis 工具对已确证的小分子目标代谢物与代谢通路的相关性进行分析发现，差异性化合物参与机体内相应代谢途径。

腺苷是一种内源性嘌呤核苷酸，具有参与调控核苷酸合成、细胞能量代谢和信号转导等多种生物学功能。[268,269] 5-羟吲哚乙醛是 5-羟色胺（5-HT，5-hydroxytryptamine）在单胺氧化酶作用下的衍生产物。[207] 5-羟色胺参与糖尿病的色氨酸代谢调节，5-羟吲哚乙醛的浓度与色氨酸代谢通路有关。[205] 鞘氨醇可参与氨基酸与其衍生物的代谢调节。茶碱在葡萄糖、蔗糖、胆盐、有机酸与胺类通路调节中具有重要作用。牛磺胆酸（TCA）与甘氨胆酸（GCA）是胆酸代谢的产物，胆酸可能在脂肪酸生物合成、糖酵解、脂质与脂蛋白代谢中具有重要作用。[208] 前期研究发现，植物乳杆菌 NCU116 和发酵胡萝卜汁表现出缓解脂代谢紊乱的特性，与本研究二者能够影响血清 TCA 和 GCA 浓度结果具有一定的相关性。[265]

表9-2 正离子模式下潜在生物标记物的鉴别与相关通路

Table 9-2 Potential biomarkers identified in positive ion mode and the related pathways

Rt_[M+H]+	Actual_M	Proposed compound	MS/MS	Proposed structure	potential pathway involved
1.36_268.1043	267.2413	Adenosine	222.0995 $C_9H_{12}N_5O_2$ 138.0557 $C_7H_8NO_2$ 138.0557 $C_5H_6N_4O$ 136.0645 $C_7H_8N_2O$ 90.0550 $C_3H_8NO_2$ 69.0362 $C_2H_3N_3$		Drug induction of bile acid pathway
1.40_116.0705	115.1305	Proline	70.0650 C_4H_8N 68.0475 C_4H_6N 59.0731 C_3H_9N 58.0645 C_3H_8N		Tryptophan metabolism Transport of glucose and other sugars, bile salts and organic acids, metal ions and amine compounds
10.81_176.0707	175.1840	5-Hydroxy indole acetaldehyde	161.0589 $C_{10}H_9O_2$ 105.0342 C_7H_5O 104.0480 C_7H_6N 89.0590 $C_4H_9O_2$		Tryptophan metabolism

续表

Rt_[M+H]+	Actual_M	Proposed compound	MS/MS	Proposed structure	potential pathway involved
12.30_466.3160	465.6227	Glycocholic acid	282.1659 $C_{15}H_{24}NO_4$ 185.0930 $C_{13}H_{13}O$ 173.1399 $C_9H_{19}NO_2$ 156.0908 $C_{12}H_{12}$ 105.0675 C_8H_9 76.0392 $C_2H_6NO_2$ 55.0540 C_4H_7		Fatty acid biosynthesis, glycolysis and gluconeogenesis, metabolism of lipids and lipoproteins
13.68_500.3041	499.7036	Taurochenodeoxycholic acid	344.2616 $C_{19}H_{38}NO_2S$ 343.2638 $C_{20}H_{39}O_2S$ 227.1847 $C_{13}H_{25}NO_2$ 201.1646 $C_{15}H_{21}$ 93.0698 C_7H_9 82.0759 C_6H_{10}		GPCR downstream signaling
17.59_300.2904	299.4919	Sphingosine	282.2772 $C_{18}H_{36}NO$ 257.2443 $C_{16}H_{33}O_2$ 155.1794 $C_{11}H_{23}$ 85.0979 C_6H_{13} 71.0862 C_5H_{11} 62.0612 C_2H_8NO		Metabolism of amino acids and derivatives

表9-3 负离子模式下潜在生物标记物的鉴别与相关通路

Table 9.3 Potential biomarkers identified in negative ion mode and the related pathways

Rt_[M-H]-	Actual_M	Proposed compound	MS/MS	Proposed structure	potential pathway involved
1.70_179.0569	180.1640	Theophylline	119.0477 $C_6H_5N_3$ 93.0329 $C_4H_3N_3$ 80.0423 $C_4H_4N_2$ 55.0190 C_3H_3O		Transport of glucose and other sugars, bile salts and organic acids, metal ions and amine compounds
12.79_514.2847	515.7030	Taurocholic acid	400.2865 $C_{25}H_{38}NO_3$ 389.2960 $C_{24}H_{39}NO_3$ 152.0034 $C_3H_6NO_4S$ 136.0126 $C_3H_6NO_3S$ 124.0096 $C_2H_6NO_3S$ 80.9634 HO_3S		Glycolysis and gluconeogenesis, triacylglyceride synthesis

本实验利用 UPLC-Q-TOF/MS 来研究植物乳杆菌 NCU116 及其发酵胡萝卜汁对糖尿病大鼠在生物学层面的影响。不同生物标记物的浓度经植物乳杆菌 NCU116 及发酵胡萝卜汁干预 5 周后发生变化，结果提示植物乳杆菌 NCU116 及发酵胡萝卜汁可能具有缓解糖尿病症状的作用。另外，植物乳杆菌 NCU116 组和发酵胡萝卜汁组含有相同浓度的植物乳杆菌 NCU116，并且两组在统计分析时表现得更为接近（图 9-2 和图 9-3）。因此，表明植物乳杆菌 NCU116 及发酵胡萝卜汁表现出的缓解糖尿病症状的功能与植物乳杆菌 NCU116 的补充有关。

9.5　本章小结

本章通过高脂高糖喂养结合小剂量 STZ 诱导来建立 II 型糖尿病大鼠模型，利用 UPLC-Q-TOF/MS 来研究植物乳杆菌 NCU116 及发酵胡萝卜汁对糖尿病血清代谢组学的影响。通过非配对 t 检验，PCA 和 PLS-DA 分析可知，植物乳杆菌 NCU116 及发酵胡萝卜汁对糖尿病大鼠血清中对腺苷（adenosine）、5-羟吲哚乙醛（5-hydroxy indole acetaldehyde）、脯氨酸（proline）、甘氨胆酸（glycocholic acid）、牛磺鹅去氧胆酸（taurochenodeoxycholic acid）、鞘氨醇（sphingosine）、茶碱（theophylline）和牛磺胆酸（taurocholic acid）代谢均有影响，并通过调节机体内葡萄糖、脂肪酸和脂蛋白等代谢途径，来发挥抗糖尿病作用。另外，植物乳杆菌 NCU116 及发酵胡萝卜汁表现出的缓解糖尿病症状的功能与植物乳杆菌 NCU116 的补充有关。

第10章 结论与展望

10.1 本研究的主要结论

该书以本实验从泡菜中筛选得到的一株体外性能良好的植物乳杆菌NCU116为研究对象,通过建立肠道菌群调节、便秘、高脂血症、结肠炎和糖尿病等动物模型对该菌的体内活性进行综合评价,并探讨其可能的益生特性。现将本书所取得的主要结果和结论总结如下:

(1)通过小鼠灌胃不同剂量植物乳杆菌NCU116,评估该益生菌对肠道微生态平衡的作用。结果表明,植物乳杆菌NCU116能够促进乳酸杆菌、双歧杆菌的生长和抑制肠杆菌、肠球菌生长。另外,该菌在产生短链脂肪酸、降低血脂水平、抑制氧化应激和调节血清细胞因子方面具有一定的作用。该研究证实,植物乳杆菌NCU116具有一定的肠道菌群调节作用。

(2)利用洛哌丁胺诱导产生小鼠便秘模型,研究植物乳杆菌NCU116对便秘症状的缓解作用。结果表明,该菌对便秘小鼠的粪便指标,肠道蠕动性能具有显著的改善作用;该菌能够促进结肠短链脂肪酸的产生,并在组织病理学研究上促进结肠上皮细胞与黏膜损伤的修复;该菌还能够改善胃肠道间质细胞(ICC)的 c-kit 基因表达,从而增加结肠的蠕动性能,促进排便。该研究证实,植物乳杆菌NCU116对便秘症状具有一定的缓解能力。

(3)利用高脂高胆固醇类饲料喂养大鼠建立脂代谢紊乱动物模型,给予大鼠灌胃不同剂量植物乳杆菌NCU116进行干预。结果表明,植物乳杆菌NCU116对高脂饮食诱导高脂血症大鼠模型具有潜在的调节血脂、降低氧化应激、缓解胰腺和脂肪病变的功效;该菌还具有调控脂代谢与胆固醇代谢基因表达的作用。该研究证实,植物乳杆菌NCU116对高脂饮食诱导的高脂血症大鼠症状具有一定的抑制作用,提示该菌对以高脂血症为代表的代谢综合征的预防有着重要意义。

（4）高脂饮食具有促进非酒精性脂肪肝症状产生的性能。植物乳杆菌NCU116在干预非酒精性脂肪肝病中，表现出修复肝功能和调节结肠菌群的作用，并且该菌在降低肝脏脂肪积累，改善脂肪肝的微观结构，促进肝损伤修复、降低血清脂多糖和促炎因子，调节脂代谢相关基因表达等方面具有较好作用。植物乳杆菌NCU116对肝脏脂肪代谢的调节可能是通过下调脂肪合成基因表达和上调脂肪氧化基因表达两条通路来实现的。

（5）取高脂饮食大鼠血清，利用UPLC-Q-TOF/MS检测植物乳杆菌NCU116对高脂饮食大鼠代谢组学的影响。结果表明，在正、负离子模式下可鉴别出13种不同的生物标记物。植物乳杆菌NCU116可通过这些生物标记物来调节机体内脂质、葡萄糖和脂蛋白等代谢途径，来发挥抑制高脂血症作用；植物乳杆菌NCU116表现出的缓解高脂血症症状的功能与植物乳杆菌NCU116高剂量补充有关。

（6）通过三硝基苯磺酸诱导建立小鼠结肠炎模型，给予植物乳杆菌NCU116及发酵胡萝卜汁干预后，发现结肠炎小鼠粪便短链脂肪酸水平显著升高、氧化应激和促炎因子水平降低、结肠黏膜损伤明显降低，其可能的机制与短链脂肪酸的含量有关。结果表明，植物乳杆菌NCU116及发酵胡萝卜汁对结肠炎小鼠结肠损伤具有缓解作用。通过对比可知，该菌及发酵胡萝卜汁对结肠炎的缓解作用与植物乳杆菌NCU116的供给密切相关。

（7）研究利用高脂高糖饮食结合小剂量链脲佐菌素诱导成功建立Ⅱ型糖尿病大鼠模型，给予植物乳杆菌NCU116及发酵胡萝卜汁干预后，糖尿病大鼠血糖、血脂、激素与氧化应激、短链脂肪酸水平、胰腺组织损伤、糖代谢与脂代谢基因表达水平明显改善。初步证实，植物乳杆菌NCU116及发酵胡萝卜汁具有缓解糖尿病症状的功能。

（8）取糖尿病大鼠血清，采用UPLC-Q-TOF/MS来分析植物乳杆菌NCU116和发酵胡萝卜汁在Ⅱ型糖尿病大鼠模型中的代谢物与通路。结果表明，通过非配对t检验，PCA和PLS-DA分析可知，植物乳杆菌NCU116及发酵胡萝卜汁对糖尿病大鼠血清中对8种小分子化合物代谢均有影响，并通过调节机体内葡萄糖、脂肪酸和脂蛋白等代谢途径，来发挥抗糖尿病作用。此外，植物乳杆菌NCU116及发酵胡萝卜汁表现出的缓解糖尿病症状的功能与植物乳杆菌NCU116的补充有关。

10.2 进一步的研究方向

虽然本研究发现植物乳杆菌 NCU116（及发酵胡萝卜汁）在肠道菌群调节、降低胆固醇、抑制结肠炎、降血糖等方面具有一定作用。但在以下几个方面仍有待进一步研究：

（1）在肠道菌群的研究中，可结合变性梯度凝胶电泳（DGGE）和温度梯度凝胶电泳（TGGE）等技术更加全面地分析肠道菌群的种类和数量。

（2）益生菌的免疫性能是其功能活性中重要的一个方面，未来可对植物乳杆菌 NCU116 在肠道免疫方面进行深入探讨。

（3）可在代谢组学范畴内结合核磁共振（NMR）、气相色谱质谱（GC/MS）与液相色谱质谱（LC/MS）联用技术对血清或尿液代谢产物进行深入研究，更加精确地找到植物乳杆菌 NCU116 在体内代谢标记物并进行通路分析。

参 考 文 献

[1] Lilly D M, Stillwell R. H. Probiotics: growth-promoting factors produced by microorganisms [J]. Science, 1965, 147 (3659): 747-748.

[2] Afrc R. F. Probiotics in man and animals [J]. Journal of Applied Microbiology, 1989, 66 (5): 365-378.

[3] Group F. W. W. Guidelines for the evaluation of probiotics in food [M]. FAO/WHO, London, ON. 2002.

[4] Gregoret V., Perezlindo M., Vinderola G., et al. A comprehensive approach to determine the probiotic potential of human-derived *Lactobacillus* for industrial use [J]. Food Microbiology, 2013, 34 (1): 19-28.

[5] Heintz C., Mair W. You are what you host: Microbiome modulation of the aging process [J]. Cell, 2014, 156 (3): 408-411.

[6] Salminen S. J. Adhesion of some probiotic and dairy *Lactobacillus* strains to Caco-2 cell cultures [J]. International Journal of Food Microbiology, 1998, 41 (1): 45-51.

[7] Van Tassell M. L., Miller M. J. Lactobacillus adhesion to mucus [J]. Nutrients, 2011, 3 (5): 613-636.

[8] Neal-Mckinney J. M., Lu X., Duong T., et al. Production of organic acids by probiotic lactobacilli can be used to reduce pathogen load in poultry [J]. PloS One, 2012, 7 (9).

[9] Palomar M. M., Maldonado Galdeano C., Perdigón G. Influence of a probiotic lactobacillus strain on the intestinal ecosystem in a stress model mouse [J]. Brain, Behavior, and Immunity, 2014, 35 77-85.

[10] Van Baarlen P., Wells J. M., Kleerebezem M. Regulation of intestinal homeostasis and immunity with probiotic lactobacilli [J]. Trends in Immunology, 2013, 34 (5): 208-215.

[11] Kim Y., Yoon S., Lee S. B., et al. Fermentation of Soy Milk via Lactobacillus plantarum Improves Dysregulated Lipid Metabolism in Rats on a High

Cholesterol Diet [J]. PloS One, 2014, 9 (2).

[12] Chen Z.-Y., Ma K. Y., Liang Y., et al. Role and classification of cholesterol-lowering functional foods [J]. Journal of Functional Foods, 2011, 3 (2): 61-69.

[13] Von Wright A., Axelsson L. Lactic acid bacteria: an introduction [M]. Lahtinen S., Ouwehand A. C., Salminen S., et al. Lactic Acid Bacteria: Microbiological and Functional Aspects, CRC Press, London. 2011: 1-17.

[14] De Vries M. C., Vaughan E. E., Kleerebezem M., et al. Lactobacillus plantarum—survival, functional and potential probiotic properties in the human intestinal tract [J]. International Dairy Journal, 2006, 16 (9): 1018-1028.

[15] Hammes W. P., Vogel R. F. The genus lactobacillus [M]. The genera of lactic acid bacteria. Springer, 1995: 19-54.

[16] Rigobelo E. C. Probiotics [M]. InTech, 2012.

[17] vila C., Carvalho B., Pinto J., et al. The use of *Lactobacillus* species as starter cultures for enhancing the quality of sugar cane silage [J]. Journal of Dairy Science, 2014, 97 (2): 940-951.

[18] Saha B. C., Racine F. M. Biotechnological production of mannitol and its applications [J]. Applied Microbiology and Biotechnology, 2011, 89 (4): 879-891.

[19] Axelsson L. Lactic acid bacteria: Classification and physiology [M]. Salminen S., Von Wright A., Ouwehand A. Lactic acid bacteria: microbiology and functional aspects. Marcel Dekker, 2004: 1-66.

[20] Mcneil N. I., Cummings J., James W. Short chain fatty acid absorption by the human large intestine [J]. Gut, 1978, 19 (9): 819-822.

[21] Scheppach W. Effects of short chain fatty acids on gut morphology and function [J]. Gut, 1994, 35 (Suppl 1): S35-S38.

[22] Zhang G., Hamaker B. R. Review: cereal carbohydrates and colon health [J]. Cereal Chemistry, 2010, 87 (4): 331-341.

[23] De Baere S., Eeckhaut V., Steppe M., et al. Development of a HPLC–UV method for the quantitative determination of four short-chain fatty acids and lactic acid produced by intestinal bacteria during in vitro fermentation [J]. Journal of Pharmaceutical and Biomedical Analysis, 2013, (80)107-115.

[24] 张珍，李波清．乳酸菌主要代谢产物及其作用研究进展 [J]．滨州医学院学报，

2012, (04): 274-276.

[25] Russell D., Ross R., Fitzgerald G., *et al.* Metabolic activities and probiotic potential of bifidobacteria [J]. International Journal of Food Microbiology, 2011, 149 (1): 88-105.

[26] Van Nieuwenhove C. P., Terán V., González S. N. Conjugated linoleic and linolenic acid production by bacteria: development of functional foods [M]. Rigobelo E. C. Probiotics. Rijeka, Croatia; InTech. 2012: 55-80.

[27] Komatsuzaki N., Shima J., Kawamoto S., *et al.* Production of γ-aminobutyric acid (GABA) by *Lactobacillus paracasei* isolated from traditional fermented foods [J]. Food Microbiology, 2005, 22 (6): 497-504.

[28] 杨胜远, 陆兆新, 吕风霞, 等. γ-氨基丁酸的生理功能和研究开发进展 [J]. 食品科学, 2005, (09): 528-533.

[29] 林亲录, 王婧, 陈海军. γ-氨基丁酸的研究进展 [J]. 现代食品科技, 2008, (05): 496-500.

[30] Badel S., Bernardi T., Michaud P. New perspectives for Lactobacilli exopolysaccharides [J]. Biotechnology Advances, 2011, 29 (1): 54-66.

[31] Journal of Functional Foodisjournal of Functional Foodisliu C. F., Tseng K. C., Chiang S. S., *et al.* Immunomodulatory and antioxidant potential of Lactobacillus exopolysaccharides [J]. Journal of the Science of Food and Agriculture, 2011, 91 (12): 2284-2291.

[32] 权自芳, 叶泥, 颜其贵. 乳酸菌素研究进展及应用 [J]. 微生物学杂志, 2013, (02): 89-92.

[33] 蒋志国, 杜琪珍. 乳酸菌素研究进展 [J]. 中国酿造, 2008, (18): 1-4.

[34] Jensen H., Roos S., Jonsson H., *et al.* The role of *Lactobacillus reuteri* Cell and Mucus Binding protein A (CmbA) in adhesion to intestinal epithelial cells and mucus in vitro [J]. Microbiology, 2014, (160): 671-681.

[35] Guarner F., Malagelada J. R. Gut flora in health and disease [J]. The Lancet, 2003, 361 (9356): 512-519.

[36] 张玢. 植物乳杆菌 KF5 的筛选及其降胆固醇特性的初步研究 [D]. 天津: 天津科技大学, 2009.

[37] Valenzuela A. S., Ruiz G. D., Omar N. B., *et al.* Inhibition of food poisoning and

pathogenic bacteria by *Lactobacillus plantarum* strain 2.9 isolated from ben saalga, both in a culture medium and in food [J]. Food Control, 2008, 19 (9): 842-848.

[38] Applegate J. A., Walker C. L. F., Ambikapathi R., *et al.* Systematic review of probiotics for the treatment of community-acquired acute diarrhea in children [J]. BMC Public Health, 2013, 13 (Suppl 3): S16.

[39] Jafari S. A., Ahanchian H., Kiani M. A., *et al.* Synbiotic for prevention of antibiotic-associated diarrhea in children: A randomized clinical trial [J]. International Journal of Pediatrics, 2014, 2 (1): 43-48.

[40] Goldenberg J. Z., Ma S., Saxton J. D., *et al.* Probiotics for the prevention of Clostridium difficile-associated diarrhea in adults and children [J]. Cochrane Database Syst Rev, 2013, 5

[41] Surawicz C. M., Damman C. The role of probiotics in prevention and treatment of GI infections [M]. Intestinal Microbiota in Health and Disease: Modern Concepts, 2014: 165.

[42] 许栋明. 益生菌预防和治疗腹泻的临床效果 [J]. 中国药学杂志, 2010, (14): 1109-1111.

[43] Preidis G. A., Hill C., Guerrant R. L., *et al.* Probiotics, enteric and diarrheal diseases, and global health [J]. Gastroenterology, 2011, 140 (1): 8.

[44] Nagashima K., Yasokawa D., Abe K., *et al.* Effect of a Lactobacillus species on incidence of diarrhea in calves and change of the microflora associated with growth [J]. Bioscience and Microflora, 2010, 29 (2): 97-110.

[45] Szymański H., Armańska M., Kowalska-Duplaga K., *et al.* Bifidobacterium longum PL03, *Lactobacillus rhamnosus* KL53A, and *Lactobacillus plantarum* PL02 in the prevention of antibiotic-associated diarrhea in children: a randomized controlled pilot trial [J]. Digestion, 2008, 78 (1): 13-17.

[46] Lembo A., Camilleri M. Chronic constipation [J]. New England Journal of Medicine, 2003, 349 (14): 1360-1368.

[47] Liem O., Benninga M. A. The role of a probiotics mixture in the treatment of childhood constipation: a pilot study [J]. Nutrition Journal, 2007, (6)17.

[48] Borchers A. T., Selmi C., Meyers F. J., *et al.* Probiotics and immunity [J]. Journal of Gastroenterology, 2009, 44 (1): 26-46.

[49] 蔡凯凯，黄占旺，叶德军，等. 益生菌调节肠道菌群及免疫调节作用机理 [J]. 中国饲料，2011，(18): 34-37.

[50] 孙进，乐国伟，侯丽霞，等. 一株植物乳杆菌内化于小鼠回肠派伊尔结及其免疫调节作用的研究 [J]. 免疫学杂志，2008，(01): 49-52.

[51] Williams N. T. Probiotics [J]. American Journal of Health-System Pharmacy, 2010，67 (6): 449-458.

[52] Pradeep K., Kuttappa M., Prasana K. Probiotics and oral health: an update [J]. South African Dental Journal, 2014，69 (1): 20-24.

[53] 周晓梅. 益生菌与儿童口腔健康的关系 [J]. 社区医学杂志，2011，(05): 14-15.

[54] 杨颖，陈卫，张灏，等. 植物乳杆菌 HO-69 的口腔益生性质研究 [J]. 华西口腔医学杂志，2008，(05): 482-485+489.

[55] Lesbros-Pantoflickova D., Corthésy-Theulaz I., Blum A. L. Helicobacter pylori and probiotics [J]. The Journal of nutrition, 2007，137 (3): 812S-818S.

[56] Gotteland M., Brunser O., Cruchet S. Systematic review: are probiotics useful in controlling gastric colonization by Helicobacter pylori? [J]. Alimentary Pharmacology and Therapeutics, 2006，23 (8): 1077-1086.

[57] Rokka S., Pihlanto A., Korhonen H., *et al*. In vitro growth inhibition of Helicobacter pylori by lactobacilli belonging to the *Lactobacillus plantarum* group [J]. Letters in Applied Microbiology, 2006，43 (5): 508-513.

[58] Hoveyda N., Heneghan C., Mahtani K. R., *et al*. A systematic review and meta-analysis: probiotics in the treatment of irritable bowel syndrome [J]. BMC Gastroenterology, 2009，9 (1): 15.

[59] Law M., Wald N., Wu T., *et al*. Systematic underestimation of association between serum cholesterol concentration and ischaemic heart disease in observational studies: data from the BUPA study [J]. BMJ, 1994，308 (6925): 363-366.

[60] Hu X., Wang T., Li W., *et al*. Effects of NS lactobacillus strains on lipid metabolism of rats fed a high-cholesterol diet [J]. Lipids in Health and Disease, 2013，12 (1): 67.

[61] Li C., Nie S.-P., Ding Q., *et al*. Cholesterol-lowering effect of *Lactobacillus plantarum* NCU116 in a hyperlipidaemic rat model [J]. Journal of Functional Foods,

2014, (8) 340-347.

[62] Nguyen T., Kang J., Lee M. Characterization of *Lactobacillus plantarum* PH04, a potential probiotic bacterium with cholesterol-lowering effects [J]. International Journal of Food Microbiology, 2007, 113 (3): 358-361.

[63] Gilliland S., Nelson C., Maxwell C. Assimilation of cholesterol by *Lactobacillus acidophilus* [J]. Applied and Environmental Microbiology, 1985, 49 (2): 377-381.

[64] De Smet I., De Boever P., Verstraete W. Cholesterol lowering in pigs through enhanced bacterial bile salt hydrolase activity [J]. British Journal of Nutrition, 1998, 79 (2): 185-194.

[65] Pereira D. I. A., Mccartney A. L., Gibson G. R. An in vitro study of the probiotic potential of a bile-salt-hydrolyzing *Lactobacillus fermentum* strain, and determination of its cholesterol-lowering properties [J]. Applied and Environmental Microbiology, 2003, 69 (8): 4743-4752.

[66] 岳喜庆, 胡梦坤. 植物乳杆菌LP1103降低小鼠血清胆固醇作用机理的探讨 [J]. 食品科技, 2007, (12): 228-231.

[67] Mozzi F., Raya R. R., Vignolo G. M. Biotechnology of lactic acid bacteria: novel applications [M]. John Wiley & Sons, 2010.

[68] Machado M. V., Cortez-Pinto H. Gut microbiota and nonalcoholic fatty liver disease [J]. Annals of Hepatology, 2012, 11 (4): 440-449.

[69] Britton R. A., Irwin R., Quach D., *et al.* Probiotic *L. reuteri* Treatment Prevents Bone Loss in a Menopausal Ovariectomized Mouse Model [J]. Journal of Cellular Physiology, 2014, 229 (11): 1822-1830.

[70] Zhao Y., Zhao L., Zheng X., *et al. Lactobacillus salivarius* strain FDB89 induced longevity in Caenorhabditis elegans by dietary restriction [J]. Journal of Microbiology, 2013, 51 (2): 183-188.

[71] Bajaj J., Heuman D., Hylemon P., *et al.* Randomised clinical trial: *Lactobacillus* GG modulates gut microbiome, metabolome and endotoxemia in patients with cirrhosis [J]. Alimentary Pharmacology and Therapeutics, 2014, 39 (10): 1113-1125.

[72] Naidu A., Bidlack W., Clemens R. Probiotic spectra of lactic acid bacteria (LAB) [J]. Critical Reviews in Food Science and Nutrition, 1999, 39 (1): 13-126.

[73] 刘中辉，彭俊生. 益生菌临床应用新进展 [J]. 肠外与肠内营养，2007，(05)：313-316,318.

[74] Pavan S., Desreumaux P., Mercenier A. Use of mouse models to evaluate the persistence, safety, and immune modulation capacities of lactic acid bacteria [J]. Clinical and Diagnostic Laboratory Immunology, 2003，10 (4): 696-701.

[75] Adawi D., Molin G., Ahrné S., et al. Safety of the probiotic strain *Lactobacillus plantarum* DSM 9843 (= strain 299v) in an endocarditis animal model [J]. Microbial Ecology in Health and Disease, 2002，14 (1): 50-53.

[76] Kneifel W., Salminen S. Probiotics and health claims [M]. Wiley Online Library, 2011.

[77] Heller K. J., Bockelmann W., Schrezenmeir J. Cheese and Its Potential as a Probiotic Food [M]. Handbook of fermented functional foods,2008: 243-267.

[78] Lourens-Hattingh A., Viljoen B. C. Yogurt as probiotic carrier food [J]. International Dairy Journal, 2001，11 (1): 1-17.

[79] Casalta E., Cervi M., Salmon J., et al. White wine fermentation: interaction of assimilable nitrogen and grape solids [J]. Australian Journal of Grape and Wine Research, 2013，19 (1): 47-52.

[80] 黄持都，鲁绯，纪凤娣，等. 酱油研究进展 [J]. 中国酿造，2009，(10): 7-9,28.

[81] Adams M. Vinegar [M]. Microbiology of fermented foods. Springer,1997: 1-44.

[82] Xiong T., Guan Q. Q., Song S. H., et al. Dynamic changes of lactic acid bacteria flora during Chinese sauerkraut fermentation [J]. Food Control, 2012，26 (1): 178-181.

[83] Farnworth E. R., Mainville I. Kefir: a fermented milk product [M]. Handbook of fermented functional foods. 2003: 77-111.

[84] Fonteles T. V., Costa M. G. M., De Jesus A. L. T., et al. Optimization of the fermentation of cantaloupe juice by *Lactobacillus casei* NRRL B-442 [J]. Food and Bioprocess Technology, 2012，5 (7): 2819-2826.

[85] Kun S., Rezessy-Szabó J. M., Nguyen Q. D., et al. Changes of microbial population and some components in carrot juice during fermentation with selected *Bifidobacterium* strains [J]. Process Biochemistry, 2008，43 (8): 816-821.

[86] Burton G. W., Ingold K. U. Beta-carotene: an unusual type of lipid antioxidant [J].

Science, 1984, 224 (4649): 569-573.

[87] Maritim A., Dene B. A., Sanders R. A., *et al.* Effects of β - carotene on oxidative stress in normal and diabetic rats [J]. Journal of Biochemical and Molecular Toxicology, 2002, 16 (4): 203-208.

[88] Omenn G. S., Goodman G. E., Thornquist M. D., *et al.* Effects of a combination of beta carotene and vitamin A on lung cancer and cardiovascular disease [J]. New England Journal of Medicine, 1996, 334 (18): 1150-1155.

[89] Nicolle C., Gueux E., Jaffrelo L., *et al.* Lyophilized carrot ingestion lowers lipemia and beneficially affects cholesterol metabolism in cholesterol - fed C57BL/6J mice [J]. European Journal of Nutrition, 2004, 43 (4): 237-245.

[90] Demir N., Bah ecİ K. S., Acar J. The effects of different initial *Lactobacillus plantarum* concentrations on some properties of fermented carrot juice [J]. Journal of Food Processing and Preservation, 2006, 30 (3): 352-363.

[91] 马晓娟. 乳酸菌发酵胡萝卜原浆技术及其产品性能研究 [D]. 南昌大学, 2013.

[92] 孙茂成, 李艾黎, 霍贵成, 等. 乳酸菌代谢组学研究进展 [J]. 微生物学通报, 2012, (10): 1499-1505.

[93] 夏建飞, 梁琼麟, 胡坪, 等. 代谢组学研究策略与方法的新进展 [J]. 分析化学, 2009, (01): 136-143.

[94] Lu X., Zhao X., Bai C., *et al.* LC - MS-based metabonomics analysis [J]. Journal of Chromatography B, 2008, 866 (1): 64-76.

[95] Zhang A., Sun H., Wang P., *et al.* Modern analytical techniques in metabolomics analysis [J]. Analyst, 2012, 137 (2): 293-300.

[96] Zhang A.-H., Qiu S., Xu H.-Y., *et al.* Metabolomics in diabetes [J]. Clinica Chimica Acta, 2014, (429) : 106-110.

[97] Gika H. G., Theodoridis G. A., Plumb R. S., *et al.* Current practice of liquid chromatography - mass spectrometry in metabolomics and metabonomics [J]. Journal of Pharmaceutical and Biomedical Analysis, 2014, (87) : 12-25.

[98] Xiong T., Huang X. H., Huang J. Q., *et al.* High-density cultivation of *Lactobacillus plantarum* NCU 116 in an ammonium and glucose fed-batch system [J]. African Journal of Biotechnology, 2011, 10 (38): 7518-7525.

[99] 熊涛, 宋苏华, 黄锦卿, 等. 植物乳杆菌 NCU116 在模拟人体消化环境中的

耐受力 [J]. 食品科学，2011，(11): 114-117.

[100] 熊涛，宋苏华，黄涛，等. 植物乳杆菌 NCU116 抑菌性能的研究 [J]. 食品与发酵工业，2012，(06): 97-101.

[101] Xiong T., Song S., Huang X., *et al.* Screening and identification of functional *Lactobacillus* specific for vegetable fermentation [J]. Journal of Food Science, 2013，78 (1): 84-89.

[102] Tancrede C. Role of human microflora in health and disease [J]. European Journal of Clinical Microbiology and Infectious Diseases, 1992，11 (11): 1012-1015.

[103] Bäckhed F., Ding H., Wang T., *et al.* The gut microbiota as an environmental factor that regulates fat storage [J]. Proceedings of the National Academy of Sciences of the United States of America, 2004，101 (44): 15718-15723.

[104] Sekirov I., Russell S. L., Antunes L. C. M., *et al.* Gut microbiota in health and disease [J]. Physiological Reviews, 2010，90 (3): 859-904.

[105] 范薇. 实验小鼠肠道正常菌群 [J]. 中国比较医学杂志，2004，(01): 57-59.

[106] Wong J. M., De Souza R., Kendall C. W., *et al.* Colonic health: fermentation and short chain fatty acids [J]. Journal of Clinical Gastroenterology, 2006，40 (3): 235-243.

[107] Fooks L., Gibson G. Probiotics as modulators of the gut flora [J]. British Journal of Nutrition, 2002，88 (S1): s39-s49.

[108] Fung W.-Y., Lye H.-S., Lim T.-J., *et al.* Roles of Probiotic on Gut Health [M]. Probiotics. Springer, 2011: 139-165.

[109] Sanz Y., Nadal I., Sanchez E. Probiotics as drugs against human gastrointestinal infections [J]. Recent patents on anti-infective drug discovery, 2007，2 (2): 148-156.

[110] Ng S., Hart A., Kamm M., *et al.* Mechanisms of action of probiotics: recent advances [J]. Inflammatory Bowel Diseases, 2009，15 (2): 300-310.

[111] Hemarajata P., Versalovic J. Effects of probiotics on gut microbiota: mechanisms of intestinal immunomodulation and neuromodulation [J]. Therapeutic Advances in Gastroenterology, 2012，1756283X12459294.

[112] Stolaki M., De Vos W. M., Kleerebezem M., *et al.* Lactic acid bacteria in the gut [M]. Lahtinen S., Ouwehand A. C., Salminen, et al. Lactic acid bacteria: microbiological

and functional aspects New York; CRC Press. 2011: 385-402.

[113] Molin G. Probiotics in foods not containing milk or milk constituents, with special reference to *Lactobacillus plantarum* 299v [J]. The American journal of clinical nutrition, 2001, 73 (2): 380s-385s.

[114] Yanbo W., Zirong X. Effect of probiotics for common carp (Cyprinus carpio) based on growth performance and digestive enzyme activities [J]. Animal Feed Science and Technology, 2006, 127 (3): 283-292.

[115] Tannock G., Munro K., Harmsen H., et al. Analysis of the fecal microflora of human subjects consuming a probiotic product containing Lactobacillus rhamnosus DR20 [J]. Applied and Environmental Microbiology, 2000, 66 (6): 2578-2588.

[116] Liong M., Shah N. Effects of a *Lactobacillus casei* synbiotic on serum lipoprotein, intestinal microflora, and organic acids in rats [J]. Journal of Dairy Science, 2006, 89 (5): 1390-1399.

[117] Hu J.-L., Nie S.-P., Xie M.-Y. High pressure homogenization increases antioxidant capacity and short-chain fatty acid yield of polysaccharide from seeds of *Plantago asiatica* L [J]. Food Chemistry, 2013, 138 (4): 2338-2345.

[118] Kullisaar T., Songisepp E., Zilmer M. Probiotics and Oxidative Stress [M]// Lushchak V. Oxidative Stress-Environmental Induction and Dietary Antioxidants. Croatia:Intech, 2012: 203-222.

[119] 彭新颜, 于海洋, 李杰, 等. 乳酸菌抗氧化作用研究进展 [J]. 食品科学, 2012(23): 370-374.

[120] Xu X., Xu P., Ma C., et al. Gut microbiota, host health, and polysaccharides [J]. Biotechnology Advances, 2013, 31 (2): 318-337.

[121] Lee H. Y., Kim J. H., Jeung H. W., et al. Effects of Ficus carica paste on loperamide-induced constipation in rats [J]. Food and Chemical Toxicology, 2012, 50 (3-4): 895-902.

[122] Koebnick C., Wagner I., Leitzmann P., et al. Probiotic beverage containing *Lactobacillus casei* Shirota improves gastrointestinal symptoms in patients with chronic constipation [J]. Canadian Journal of Gastroenterology, 2003, 17 (11): 655-660.

[123] Dimidi E., Christodoulides S., Fragkos K., et al. The effect of probiotics on

functional constipation: a systematic review of randomised controlled trials [J]. Proceedings of the Nutrition Society, 2014, 73 (OCE1): E16.

[124] Shimotoyodome A., Meguro S., Hase T., et al. Decreased colonic mucus in rats with loperamide-induced constipation [J]. Comparative Biochemistry and Physiology Part A, 2000, 126 (2): 203-212.

[125] Wedel T., Spiegler J., Soellner S., et al. Enteric nerves and interstitial cells of Cajal are altered in patients with slow-transit constipation and megacolon [J]. Gastroenterology, 2002, 123 (5): 1459-1467.

[126] Locke G. R., Pemberton J. H., Phillips S. F. AGA technical review on constipation [J]. Gastroenterology, 2000, 119 (6): 1766-1778.

[127] Mori T., Shibasaki Y., Matsumoto K., et al. Mechanisms That Underlie μ-Opioid Receptor Agonist - Induced Constipation: Differential Involvement of μ-Opioid Receptor Sites and Responsible Regions [J]. Journal of Pharmacology and Experimental Therapeutics, 2013, 347 (1): 91-99.

[128] Neri F., Cavallari G., Tsivian M., et al. Effect of colic vein ligature in rats with loperamide-induced constipation [J]. Journal of Biomedicine and Biotechnology, 2012, 2012 1-6.

[129] Pan L.-X., Fang X.-J., Yu Z., et al. Encapsulation in alginate-skim milk microspheres improves viability of *Lactobacillus bulgaricus* in stimulated gastrointestinal conditions [J]. International Journal of Food Sciences and Nutrition, 2013, 64 (3): 380-384.

[130] Qian Y., Suo H., Du M., et al. Preventive effect of *Lactobacillus fermentum* Lee on activated carbon-induced constipation in mice [J]. Experimental and Therapeutic Medicine, 2015, 9 (1): 272-278.

[131] Picard C., Fioramonti J., Francois A., et al. Review article: bifidobacteria as probiotic agents-physiological effects and clinical benefits [J]. Alimentary Pharmacology and Therapeutics, 2005, 22 (6): 495-512.

[132] Jeon J. R., Choi J. H. Lactic acid fermentation of germinated barley fiber and proliferative function of colonic epithelial cells in loperamide-induced rats [J]. Journal of Medicinal Food, 2010, 13 (4): 950-960.

[133] Wintola O. A., Sunmonu T. O., Afolayan A. J. The effect of Aloe ferox Mill.

in the treatment of loperamide-induced constipation in Wistar rats [J]. BMC Gastroenterology, 2010, 10 (1): 95.

[134] Hou M.-L., Chang L.-W., Lin C.-H., et al. Comparative pharmacokinetics of rhein in normal and loperamide-induced constipated rats and microarray analysis of drug-metabolizing genes [J]. Journal of Ethnopharmacology, 2014, 155 (2): 1291-1299.

[135] Ashafa A., Sunmonu T., Abass A., et al. Laxative potential of the ethanolic leaf extract of *Aloe vera* (L.) Burm. f. in Wistar rats with loperamide-induced constipation [J]. Journal of Natural Pharmaceuticals, 2011, 2 (3): 158-162.

[136] Niwa T., Nakao M., Hoshi S., *et al.* Effect of dietary fiber on morphine-induced constipation in rats [J]. Bioscience, Biotechnology, and Biochemistry, 2002, 66 (6): 1233-1240.

[137] Hu J. L., Nie S. P., Min F. F., *et al.* Polysaccharide from seeds of *Plantago asiatica* L. increases short-chain fatty acid production and fecal moisture along with lowering pH in mouse colon [J]. Journal of Agricultural and Food Chemistry, 2012, 60 (46): 11525-11532.

[138] Araki Y., Fujiyama Y., Andoh A., et al. The dietary combination of germinated barley foodstuff plus *Clostridium butyricum* suppresses the dextran sulfate sodium-induced experimental colitis in rats [J]. Scandinavian Journal of Gastroenterology, 2000, 35 (10): 1060-1067.

[139] Pruzzo R. C., Paola Mastrantonio, C. Short chain fatty acids, menaquinones and ubiquinones and their effects on the host [J]. Microbial Ecology in Health and Disease, 2000, 12 (2): 209-215.

[140] Topping D. L., Clifton P. M. Short-chain fatty acids and human colonic function: roles of resistant starch and nonstarch polysaccharides [J]. Physiological Reviews, 2001, 81 (3): 1031-1064.

[141] Burns A. J., Herbert T. M., Ward S. M., et al. Interstitial cells of Cajal in the guinea-pig gastrointestinal tract as revealed by c-Kit immunohistochemistry [J]. Cell and Tissue Research, 1997, 290 (1): 11-20.

[142] He C. L., Burgart L., Wang L., et al. Decreased interstitial cell of Cajal volume in patients with slow-transit constipation [J]. Gastroenterology, 2000, 118 (1): 14-21.

[143] Yarden Y., Kuang W. J., Yang-Feng T., et al. Human proto-oncogene c-*kit*: a new cell surface receptor tyrosine kinase for an unidentified ligand [J]. The EMBO Journal, 1987, 6 (11): 3341-3351.

[144] Lee J. I., Park H., Kamm M. A., et al. Decreased density of interstitial cells of Cajal and neuronal cells in patients with slow-transit constipation and acquired megacolon [J]. Journal of Gastroenterology and Hepatology, 2005, 20 (8): 1292-1298.

[145] Chen Z. Y., Jiao R., Ma K. Y. Cholesterol-lowering nutraceuticals and functional foods [J]. Journal of Agricultural and Food Chemistry, 2008, 56 (19): 8761-8773.

[146] Young V. B. The intestinal microbiota in health and disease [J]. Current opinion in gastroenterology, 2012, 28 (1): 63-69.

[147] Omar J. M., Chan Y.-M., Jones M. L., et al. *Lactobacillus fermentum* and *Lactobacillus amylovorus* as probiotics alter body adiposity and gut microflora in healthy persons [J]. Journal of Functional Foods, 2013, 5 (1): 116-123.

[148] Xing X., Zhang Z., Hu X., et al. Antidiabetic effects of *Artemisia sphaerocephala* Krasch. gum, a novel food additive in China, on streptozotocin-induced type 2 diabetic rats [J]. Journal of Ethnopharmacology, 2009, 125 (3): 410-416.

[149] Zhu K., Nie S., Li C., et al. A newly identified polysaccharide from *Ganoderma atrum* attenuates hyperglycemia and hyperlipidemia [J]. International Journal of Biological Macromolecules, 2013, 57:142-150.

[150] Parolini C., Manzini S., Busnelli M., et al. Effect of the combinations between pea proteins and soluble fibres on cholesterolaemia and cholesterol metabolism in rats [J]. British Journal of Nutrition, 2013, 110 (8): 1-8.

[151] Rigamonti E., Parolini C., Marchesi M., et al. Hypolipidemic effect of dietary pea proteins: Impact on genes regulating hepatic lipid metabolism [J]. Molecular Nutrition & Food Research, 2010, 54 (S1): S24-S30.

[152] Mandimika T., Paturi G., De Guzman C. E., et al. Effects of dietary broccoli fibre and corn oil on serum lipids, faecal bile acid excretion and hepatic gene expression in rats [J]. Food Chemistry, 2012, 131 (4): 1272-1278.

[153] Aziz M. T. A., El-Asmar M. F., Haidara M., et al. Effect of bone marrow-derived mesenchymal stem cells on cardiovascular complications in diabetic rats [J].

Medical Science Monitor, 2008, 14 (11): 249-255.

[154] Ross R. The pathogenesis of atherosclerosis: a perspective for the 1990s [J]. Nature, 1993, 362: 801-809.

[155] Guo J., Bei W., Hu Y., et al. A new TCM formula FTZ lowers serum cholesterol by regulating HMG-CoA reductase and CYP7A1 in hyperlipidemic rats [J]. Journal of Ethnopharmacology, 2011, 135 (2): 299-307.

[156] Bao Y., Wang Z., Zhang Y., et al. Effect of *Lactobacillus plantarum* P-8 on lipid metabolism in hyperlipidemic rat model [J]. European Journal of Lipid Science and Technology, 2012, 114 (11): 1230-1236.

[157] Gotto A. M., Brinton E. A. Assessing low levels of high-density lipoprotein cholesterol as a risk factor in coronary heart disease [J]. Journal of the American College of Cardiology, 2004, 43 (5): 717-724.

[158] Pischon T., Girman C. J., Sacks F. M., et al. Non-high-density lipoprotein cholesterol and apolipoprotein B in the prediction of coronary heart disease in men [J]. Circulation, 2005, 112 (22): 3375-3383.

[159] Cani P. D., Bibiloni R., Knauf C., et al. Changes in gut microbiota control metabolic endotoxemia-induced inflammation in high-fat diet-induced obesity and diabetes in mice [J]. Diabetes, 2008, 57 (6): 1470-1481.

[160] Lee M. J., Fried S. K. Multilevel regulation of leptin storage, turnover, and secretion by feeding and insulin in rat adipose tissue [J]. Journal of Lipid Research, 2006, 47 (9): 1984-1993.

[161] Hansen G., Jelsing J., Vrang N. Effects of liraglutide and sibutramine on food intake, palatability, body weight and glucose tolerance in the gubra DIO-rats [J]. Acta Pharmacologica Sinica, 2012, 33 (2): 194-200.

[162] Kadowaki T., Yamauchi T., Kubota N., et al. Adiponectin and adiponectin receptors in insulin resistance, diabetes, and the metabolic syndrome [J]. Journal of Clinical Investigation, 2006, 116 (7): 1784-1792.

[163] Tsimikas S., I Miller Y. Oxidative modification of lipoproteins: mechanisms, role in inflammation and potential clinical applications in cardiovascular disease [J]. Current Pharmaceutical Design, 2011, 17 (1): 27-37.

[164] Xin J., Zeng D., Wang H., et al. Preventing non-alcoholic fatty liver disease through

Lactobacillus johnsonii BS15 by attenuating inflammation and mitochondrial injury and improving gut environment in obese mice [J]. Applied Microbiology and Biotechnology, 2014, 98 : 6817–6829.

[165] Andziak B., O' connor T. P., Qi W., et al. High oxidative damage levels in the longest-living rodent, the naked mole-rat [J]. Aging cell, 2006, 5 (6): 463–471.

[166] Araujo F. B., Barbosa D. S., Hsin C. Y., et al. Evaluation of oxidative stress in patients with hyperlipidemia [J]. Atherosclerosis, 1995, 117 (1): 61–71.

[167] Gamboa-Gómez C., Salgado L. M., González-Gallardo A., et al. Consumption of *Ocimum sanctum* L. and *Citrus paradisi* infusions modulates lipid metabolism and insulin resistance in obese rats [J]. Food & function, 2014, 5 (5): 927–935.

[168] Niedernhofer L. J., Daniels J. S., Rouzer C. A., et al. Malondialdehyde, a product of lipid peroxidation, is mutagenic in human cells [J]. Journal of Biological Chemistry, 2003, 278 (33): 31426–31433.

[169] Satapathy S. K., Ochani M., Dancho M., et al. Galantamine alleviates inflammation and other obesity-associated complications in high-fat diet – fed mice [J]. Molecular Medicine, 2011, 17 (7–8): 599–606.

[170] Kumar M., Rakesh S., Nagpal R., et al. Probiotic *Lactobacillus rhamnosus* GG and *Aloe vera* gel improve lipid profiles in hypercholesterolemic rats [J]. Nutrition, 2013, 29 : 574–579.

[171] Goldstein J. L., Brown M. S. The LDL receptor [J]. Arteriosclerosis, Thrombosis, and Vascular Biology, 2009, 29 (4): 431–438.

[172] Brown M. S., Goldstein J. L. A receptor-mediated pathway for cholesterol homeostasis [J]. Science, 1986, 232 (4746): 34–47.

[173] Wong J., Quinn C., Brown A. SREBP-2 positively regulates transcription of the cholesterol efflux gene, ABCA1, by generating oxysterol ligands for LXR [J]. Biochemical Journal, 2006, 400 : 485–491.

[174] Spady D. K., Cuthbert J. A., Willard M., et al. Adenovirus-mediated transfer of a gene encoding cholesterol 7 alpha-hydroxylase into hamsters increases hepatic enzyme activity and reduces plasma total and low density lipoprotein cholesterol [J]. Journal of Clinical Investigation, 1995, 96 (2): 700–709.

[175] Li T., Chiang J. Y. Nuclear receptors in bile acid metabolism [J]. Drug Metabolism

Reviews, 2013, 45 (1): 145-155.

[176] Tilg H., Hotamisligil G. S. Nonalcoholic fatty liver disease: cytokine-adipokine interplay and regulation of insulin resistance [J]. Gastroenterology, 2006, 131 (3): 934-945.

[177] Kirpich I. A., Mcclain C. J. Probiotics in the treatment of the liver diseases [J]. Journal of the American College of Nutrition, 2012, 31 (1): 14-23.

[178] Byrne C. D. Fatty liver: role of inflammation and fatty acid nutrition [J]. Prostaglandins, Leukotrienes and Essential Fatty Acids (PLEFA), 2010, 82 (4): 265-271.

[179] Cheng F. K., Torres D., Harrison S. Hepatitis C and lipid metabolism, hepatic steatosis, and NAFLD: still important in the era of direct acting antiviral therapy? [J]. Journal of Viral Hepatitis, 2014, 21 (1): 1-8.

[180] Zhang S., Zheng L., Dong D., et al. Effects of flavonoids from *Rosa laevigata* Michx fruit against high-fat diet-induced non-alcoholic fatty liver disease in rats [J]. Food Chemistry, 2013, 141 (3): 2108-2116.

[181] Frazier T. H., Dibaise J. K., Mcclain C. J. Gut microbiota, intestinal permeability, obesity-induced inflammation, and liver injury [J]. Journal of Parenteral and Enteral Nutrition, 2011, 35 (suppl 5): 14S-20S.

[182] Cesaro C., Tiso A., Del Prete A., et al. Gut microbiota and probiotics in chronic liver diseases [J]. Digestive and Liver Disease, 2011, 43 (6): 431-438.

[183] Mattace Raso G., Simeoli R., Iacono A., et al. Effects of a *Lactobacillus paracasei* B21060 based synbiotic on steatosis, insulin signaling and toll-like receptor expression in rats fed a high-fat diet [J]. The Journal of Nutritional Biochemistry, 2014, 25 (1): 81-90.

[184] Ritze Y., Bárdos G., Claus A., et al. Lactobacillus rhamnosus GG Protects against Non-Alcoholic Fatty Liver Disease in Mice [J]. PloS One, 2014, 9 (1).

[185] Li C., Nie S.-P., Zhu K.-X., et al. Effect of *Lactobacillus plantarum* NCU116 on loperamide-induced constipation in mice [J]. International Journal of Food Sciences and Nutrition, 2015.

[186] Li J., Fan Y., Zhang Z., et al. Evaluating the trans fatty acid, CLA, PUFA and erucic acid diversity in human milk from five regions in China [J]. Lipids, 2009, 44 (3):

257-271.

[187] Yang M., Yang Y., Nie S., et al. Analysis and formation of trans fatty acids in corn oil during the heating process [J]. Journal of the American Oil Chemists' Society, 2012, 89 (5): 859-867.

[188] Chen Y., Yang Y., Nie S., et al. The analysis of *trans* fatty acid profiles in deep frying palm oil and chicken fillets with an improved gas chromatography method [J]. Food Control, 2014, 44 : 191-197.

[189] Higashikawa F., Noda M., Awaya T., et al. Improvement of constipation and liver function by plant-derived lactic acid bacteria: a double-blind, randomized trial [J]. Nutrition, 2010, 26 (4): 367-374.

[190] Reichold A., Brenner S. A., Spruss A., et al. *Bifidobacterium adolescentis* protects from the development of nonalcoholic steatohepatitis in a mouse model [J]. The Journal of Nutritional Biochemistry, 2013, 25 (2): 118-125.

[191] Wagnerberger S., Spruss A., Kanuri G., et al. *Lactobacillus casei* Shirota protects from fructose-induced liver steatosis: A mouse model [J]. The Journal of nutritional biochemistry, 2013, 24 (3): 531-538.

[192] Osman N., Adawi D., Ahrné S., et al. Endotoxin-and d-galactosamine-induced liver injury improved by the administration of *Lactobacillus*, *Bifidobacterium* and blueberry [J]. Digestive and Liver Disease, 2007, 39 (9): 849-856.

[193] Shen C.-L., Chen L., Wang S., et al. Effects of dietary fat levels and feeding durations on musculoskeletal health in female rats [J]. Food & function, 2014, 5 (3): 598-604.

[194] Kang J.-H., Yun S.-I., Park H.-O. Effects of *Lactobacillus gasseri* BNR17 on body weight and adipose tissue mass in diet-induced overweight rats [J]. The Journal of Microbiology, 2010, 48 (5): 712-714.

[195] Kılıç G. B. Highlights in Probiotic Research [M]// Kongo M. Lactic Acid Bacteria - R & D for Food, Health and Livestock Purposes. InTech. 2013: 243-262.

[196] Opal S. M., Depalo V. A. Anti-inflammatory cytokines [J]. Chest Journal, 2000, 117 (4): 1162-1172.

[197] Esposito E., Iacono A., Bianco G., et al. Probiotics reduce the inflammatory response induced by a high-fat diet in the liver of young rats [J]. The Journal of nutrition,

2009, 139 (5): 905-911.

[198] Cano P. G., Santacruz A., Moya, et al. Bacteroides uniformis CECT 7771 ameliorates metabolic and immunological dysfunction in mice with high-fat-diet induced obesity [J]. PloS One, 2012, 7 (7): 1-16.

[199] Lu H.-J., Tzeng T.-F., Liou S.-S., et al. Ruscogenin Ameliorates Experimental Nonalcoholic Steatohepatitis via Suppressing Lipogenesis and Inflammatory Pathway [J]. BioMed Research International, 2014, 2014: 1-10.

[200] Zhong Z., Zhang W., Du R., et al. Lactobacillus casei Zhang stimulates lipid metabolism in hypercholesterolemic rats by affecting gene expression in the liver [J]. European Journal of Lipid Science and Technology, 2012, 114 (3): 244-252.

[201] 陈华, 苗华, 田婷, 等. 代谢组学技术在高脂血症研究中的应用 [J]. 药物分析杂志, 2014, (04): 563-569.

[202] 朱科学. 黑灵芝多糖对Ⅱ型糖尿病大鼠的健康改善及其代谢组学初探 [D]. 南昌: 南昌大学, 2014.

[203] 赵春艳, 阿基业, 曹蓓, 等. 代谢组学在代谢性疾病研究中的进展 [J]. 中国临床药理学与治疗学, 2011, (04): 439-446.

[204] Uarrota V. G., Moresco R., Coelho B., et al. Metabolomics combined with chemometric tools (PCA, HCA, PLS-DA and SVM) for screening cassava (*Manihot esculenta Crantz*) roots during postharvest physiological deterioration [J]. Food Chemistry, 2014, 161: 67-78.

[205] Su Z.-H., Li S.-Q., Zou G.-A., et al. Urinary metabonomics study of anti-depressive effect of Chaihu-Shu-Gan-San on an experimental model of depression induced by chronic variable stress in rats [J]. Journal of Pharmaceutical and Biomedical Analysis, 2011, 55 (3): 533-539.

[206] Johnson S. C. Hierarchical clustering schemes [J]. Psychometrika, 1967, 32 (3): 241-254.

[207] Takahashi T., Yano M., Minami J., et al. Sarpogrelate hydrochloride, a serotonin2A receptor antagonist, reduces albuminuria in diabetic patients with early-stage diabetic nephropathy [J]. Diabetes Research and Clinical Practice, 2002, 58 (2): 123-129.

[208] Zhu K., Nie S., Gong D., et al. Effect of polysaccharide from Ganoderma atrum on

the serum metabolites of type 2 diabetic rats [J]. Food Hydrocolloids, 2014.

[209] Nie C., Han T., Zhang L., et al. Cross-sectional and dynamic change of serum metabolite profiling for Hepatitis B-related acute-on-chronic liver failure by UPLC/MS [J]. Journal of Viral Hepatitis, 2014, 21 (1): 53-63.

[210] Korzenik J. R., Podolsky D. K. Evolving knowledge and therapy of inflammatory bowel disease [J]. Nature reviews Drug discovery, 2006, 5 (3): 197-209.

[211] Bouma G., Strober W. The immunological and genetic basis of inflammatory bowel disease [J]. Nature Reviews Immunology, 2003, 3 (7): 521-533.

[212] Brajdić A., Mijandrušić-Sinčic B. Insights to the Ethiopathogenesis of the Inflammatory Bowel Disease [M]. Szabo I. Inflammatory Bowel Disease. INTECH Open Access Publisher, 2012: 3-21.

[213] Duary R. K., Bhausaheb M. A., Batish V. K., et al. Anti-inflammatory and immunomodulatory efficacy of indigenous probiotic Lactobacillus plantarum Lp91 in colitis mouse model [J]. Molecular Biology Reports, 2012, 39 (4): 4765-4775.

[214] Sheil B., Shanahan F., O'mahony L. Probiotic effects on inflammatory bowel disease [J]. The Journal of nutrition, 2007, 137 (3): 819S-824S.

[215] 李玲, 印遇龙, 阮征, 等. 低聚乳果糖对结肠炎大鼠血浆抗氧化系统的影响 [J]. 食品科学, 2011, (13): 321-324.

[216] 袁学勤, 王旭丹, 谢鸣, 等. 三硝基苯磺酸诱导 Balb/c 小鼠结肠炎的实验研究 [J]. 中国药理学通报, 2005, (06): 756-759.

[217] Parvez S., Malik K., Ah Kang S., *et al.* Probiotics and their fermented food products are beneficial for health [J]. Journal of Applied Microbiology, 2006, 100 (6): 1171-1185.

[218] Edge R., Mcgarvey D., Truscott T. The carotenoids as anti-oxidants-a review [J]. Journal of Photochemistry and Photobiology B: Biology, 1997, 41 (3): 189-200.

[219] Zhang H., Fan M., Paliyath G. Effects of Carotenoids and Retinoids on Immune-Mediated Chronic Inflammation in Inflammatory Bowel Disease [M]. Functional Foods, Nutraceuticals, and Degenerative Disease Prevention, 2011: 213-233.

[220] Li C., Ding Q., Nie S.-P., *et al.* Carrot Juice Fermented with *Lactobacillus plantarum* NCU116 Ameliorates Type 2 Diabetes in Rats [J]. Journal of Agricultural and Food Chemistry, 2014, 62 (49): 11884-11891.

[221] Lee I.-A., Bae E.-A., Lee J.-H., et al. Bifidobacterium longum HY8004 attenuates TNBS-induced colitis by inhibiting lipid peroxidation in mice [J]. Inflammation Research, 2010, 59 (5): 359-368.

[222] Mañé J., Lorén V., Pedrosa E., et al. Lactobacillus fermentum CECT 5716 prevents and reverts intestinal damage on TNBS - induced colitis in mice [J]. Inflammatory Bowel Diseases, 2009, 15 (8): 1155-1163.

[223] Watanabe T., Fujii J., Suzuki K., et al. Dysfunction of antioxidative enzymes in the trinitrobenzenesulfonic acid-induced colitis rat [J]. Pathophysiology, 1998, 5 (3): 191-198.

[224] 陈燕, 曹郁生, 刘晓华. 短链脂肪酸与肠道菌群 [J]. 江西科学, 2006, (01): 38-40, 69.

[225] Rose D. J., Demeo M. T., Keshavarzian A., et al. Influence of dietary fiber on inflammatory bowel disease and colon cancer: importance of fermentation pattern [J]. Nutrition Reviews, 2007, 65 (2): 51-62.

[226] Ristagno G., Fumagalli F., Porretta-Serapiglia C., et al. Hydroxytyrosol attenuates peripheral neuropathy in streptozotocin-induced diabetes in rats [J]. Journal of Agricultural and Food Chemistry, 2012, 60 (23): 5859-5865.

[227] Association A. D. Diagnosis and classification of diabetes mellitus [J]. Diabetes Care, 2013, 36 (Supplement 1): S67-S74.

[228] Karlsson F. H., Tremaroli V., Nookaew I., et al. Gut metagenome in European women with normal, impaired and diabetic glucose control [J]. Nature, 2013, 498 (7452): 99-103.

[229] Qin J., Li Y., Cai Z., et al. A metagenome-wide association study of gut microbiota in type 2 diabetes [J]. Nature, 2012, 490 (7418): 55-60.

[230] Krentz A. J., Bailey C. J. Oral antidiabetic agents [J]. Drugs, 2005, 65 (3): 385-411.

[231] Chen P., Zhang Q., Dang H., et al. Antidiabetic effect of *Lactobacillus casei* CCFM0412 in high-fat-fed, streptozotocin-induced type 2 diabetic mice [J]. Nutrition, 2014, 30 (9): 1061-1068.

[232] Devaraj S., Hemarajata P., Versalovic J. The human gut microbiome and body metabolism: implications for obesity and diabetes [J]. Clinical Chemistry, 2013, 59

(4): 617-628.

[233] Kootte R., Vrieze A., Holleman F., et al. The therapeutic potential of manipulating gut microbiota in obesity and type 2 diabetes mellitus [J]. Diabetes, Obesity and Metabolism, 2012, 14 (2): 112-120.

[234] Wallace T. C., Guarner F., Madsen K., et al. Human gut microbiota and its relationship to health and disease [J]. Nutrition Reviews, 2011, 69 (7): 392-403.

[235] Rupa P., Mine Y. Recent advances in the role of probiotics in human inflammation and gut health [J]. Journal of Agricultural and Food Chemistry, 2012, 60 (34): 8249-8256.

[236] Lye H.-S., Kuan C.-Y., Ewe J.-A., et al. The improvement of hypertension by probiotics: effects on cholesterol, diabetes, renin, and phytoestrogens [J]. International Journal of Molecular Sciences, 2009, 10 (9): 3755-3775.

[237] Li C., Nie S., Zhu K.-X., et al. *Lactobacillus plantarum* NCU116 improves liver function, oxidative stress and lipid metabolism in high fat diet induced non-alcoholic fatty liver disease rats [J]. Food & function, 2014, (5): 3216-3223.

[238] Marazza J. A., Leblanc J. G., De Giori G. S., et al. Soymilk fermented with Lactobacillus rhamnosus CRL981 ameliorates hyperglycemia, lipid profiles and increases antioxidant enzyme activities in diabetic mice [J]. Journal of Functional Foods, 2013, 5 (4): 1848-1853.

[239] Srinivasan K., Viswanad B., Asrat L., et al. Combination of high-fat diet-fed and low-dose streptozotocin-treated rat: a model for type 2 diabetes and pharmacological screening [J]. Pharmacological Research, 2005, 52 (4): 313-320.

[240] Puddu A., Sanguineti R., Montecucco F., et al. Evidence for the Gut Microbiota Short-Chain Fatty Acids as Key Pathophysiological Molecules Improving Diabetes [J]. Mediators of Inflammation, 2014, 2014 1-9.

[241] Jakobsdottir G., Xu J., Molin G., et al. High-fat diet reduces the formation of butyrate, but increases succinate, inflammation, liver fat and cholesterol in rats, while dietary fibre counteracts these effects [J]. PloS One, 2013, 8 (11): e80476.

[242] Mehal W. Z. The Gordian Knot of dysbiosis, obesity and NAFLD [J]. Nature Reviews Gastroenterology and Hepatology, 2013, 10 (11): 637-644.

[243] Cameron-Smith D., Collier G., O'dea K. Effect of propionate on in vivo

carbohydrate metabolism in streptozocin-induced diabetic rats [J]. Metabolism, 1994, 43 (6): 728-734.

[244] Tremaroli V., Bäckhed F. Functional interactions between the gut microbiota and host metabolism [J]. Nature, 2012, 489 (7415): 242-249.

[245] Delzenne N. M., Neyrinck A. M., Bäckhed F., et al. Targeting gut microbiota in obesity: effects of prebiotics and probiotics [J]. Nature Reviews Endocrinology, 2011, 7 (11): 639-646.

[246] Nachnani J., Bulchandani D., Nookala A., et al. Biochemical and histological effects of exendin-4 (exenatide) on the rat pancreas [J]. Diabetologia, 2010, 53 (1): 153-159.

[247] Yadav H., Jain S., Sinha P. Antidiabetic effect of probiotic dahi containing Lactobacillus acidophilus and Lactobacillus casei in high fructose fed rats [J]. Nutrition, 2007, 23 (1): 62-68.

[248] Betteridge D. J. What is oxidative stress? [J]. Metabolism, 2000, 49 (2): 3-8.

[249] Lee J.-S. Effects of soy protein and genistein on blood glucose, antioxidant enzyme activities, and lipid profile in streptozotocin-induced diabetic rats [J]. Life Sciences, 2006, 79 (16): 1578-1584.

[250] Posuwan J., Prangthip P., Leardkamolkarn V., et al. Long-term supplementation of high pigmented rice bran oil (Oryza sativa L.) on amelioration of oxidative stress and histological changes in streptozotocin-induced diabetic rats fed a high fat diet: Riceberry bran oil [J]. Food Chemistry, 2013, 138 (1): 501-508.

[251] Kataja-Tuomola M., Sundell J., Männistö S., et al. Effect of α-tocopherol and β-carotene supplementation on the incidence of type 2 diabetes [J]. Diabetologia, 2008, 51 (1): 47-53.

[252] Bejar W., Hamden K., Ben Salah R., et al. Lactobacillus plantarum TN627 significantly reduces complications of alloxan-induced diabetes in rats [J]. Anaerobe, 2013, 2(4): 4-11.

[253] Huang S., Czech M. P. The GLUT4 Glucose Transporter [J]. Cell Metabolism, 2007, 5 (4): 237-252.

[254] Kim S.-W., Park K.-Y., Kim B., et al. Lactobacillus rhamnosus GG improves insulin sensitivity and reduces adiposity in high-fat diet-fed mice through

enhancement of adiponectin production [J]. Biochemical and Biophysical Research Communications, 2013, 431 (2): 258-263.

[255] Soares F. L. P., De Oliveira Matoso R., Teixeira L. G., et al. Gluten-free diet reduces adiposity, inflammation and insulin resistance associated with the induction of PPAR-alpha and PPAR-gamma expression [J]. The Journal of Nutritional Biochemistry, 2013, 24 (6): 1105-1111.

[256] Mohamed S. Functional foods against metabolic syndrome (obesity, diabetes, hypertension and dyslipidemia) and cardiovasular disease [J]. Trends in Food Science & Technology, 2014, 35 (2): 114-128.

[257] Zhao X., Zhang Y., Meng X., et al. Effect of a traditional Chinese medicine preparation Xindi soft capsule on rat model of acute blood stasis: a urinary metabonomics study based on liquid chromatography-mass spectrometry [J]. Journal of Chromatography B, 2008, 873 (2): 151-158.

[258] Zhang A., Sun H., Han Y., et al. Exploratory urinary metabolic biomarkers and pathways using UPLC-Q-TOF-HDMS coupled with pattern recognition approach [J]. Analyst, 2012, 137 (18): 4200-4208.

[259] Jandrić Z., Roberts D., Rathor M. N., et al. Assessment of fruit juice authenticity using UPLC–QTOF MS: A metabolomics approach [J]. Food Chemistry, 2014, (148): 7-17.

[260] Griffin J., Nicholls A. Metabolomics as a functional genomic tool for understanding lipid dysfunction in diabetes, obesity and related disorders [J]. Pharmacogenomics, 2006, 7 (7): 1095-1107.

[261] Gomes A. C., Bueno A. A., De Souza R. G. M., et al. Gut microbiota, probiotics and diabetes [J]. Nutrition journal, 2014, 13 (1): 60.

[262] Lin C.-H., Lin C.-C., Shibu M. A., et al. Oral Lactobacillus reuteri GMN-32 treatment reduces blood glucose concentrations and promotes cardiac function in rats with streptozotocin-induced diabetes mellitus [J]. British Journal of Nutrition, 2014, 111 (04): 598-605.

[263] Törrönen R., Lehmusaho M., Häkkinen S., et al. Serum β-carotene response to supplementation with raw carrots, carrot juice or purified β-carotene in healthy non-smoking women [J]. Nutrition Research, 1996, 16 (4): 565-575.

[264] Zhao Y.-Y., Shen X., Cheng X.-L., et al. Urinary metabonomics study on the protective effects of ergosta-4, 6, 8 (14), 22-tetraen-3-one on chronic renal failure in rats using UPLC Q-TOF/MS and a novel MS^E data collection technique [J]. Process Biochemistry, 2012, 47 (12): 1980-1987.

[265] Li C., Ding Q., Nie S., et al. Carrot juice fermented with *Lactobacillus plantarum* NCU116 ameliorates type 2 diabetes in rats [J]. Journal of Agricultural and Food Chemistry, 2014, 62 (49): 11884-11891.

[266] Chen J., Zhao X., Fritsche J., et al. Practical approach for the identification and isomer elucidation of biomarkers detected in a metabonomic study for the discovery of individuals at risk for diabetes by integrating the chromatographic and mass spectrometric information [J]. Analytical Chemistry, 2008, 80 (4): 1280-1289.

[267] Liu Y.-T., Peng J.-B., Jia H.-M., et al. Urinary metabonomic evaluation of the therapeutic effect of traditional Chinese medicine Xin-Ke-Shu against atherosclerosis rabbits using UPLC-Q/TOF MS [J]. Chemometrics and Intelligent Laboratory Systems, 2014, (136): 104-114.

[268] Guzman-Gutierrez E., Abarzua F., Belmar C., et al. Functional link between adenosine and insulin: a hypothesis for fetoplacental vascular endothelial dysfunction in gestational diabetes [J]. Current Vascular Pharmacology, 2011, 9 (6): 750-762.

[269] Guzmán-Gutiérrez E., Arroyo P., Pardo F., et al. The adenosine-insulin signalling axis in the fetoplacental endothelial dysfunction in gestational diabetes [M]. Gestational Diabetes-Causes, Diagnosis and Treatment, 2013: 49-77.

英文缩略语

缩略语	英文名	中文名
ACC	Acetyl-coenzyme A carboxylase	乙酰辅酶 A 羧化酶
ALT	Alanine aminotransferase	谷丙转氨酶
AST	Aspartate aminotransferase	谷草转氨酶
BW	Body weight	体重
CAT	Catalase	过氧化氢酶
CFU	Colony forming unit	菌落形成单位
CPT1α	Carnitine palmitoyltransferase-1α	肉碱棕榈酰转移酶 1
CVD	Cardiac Vascular Disease	心血管疾病
CYP7A1	Cholesterol 7α-hydroxylase	胆固醇 7α-羟化酶
ESI	Electrospray ionization source	电喷雾离子源
FA	Fatty acids	脂肪酸
FAS	Fatty acid synthetase	脂肪酸合成酶
FBG	Fasting blood glucose	空腹血糖
FID	Flame ionization detector	火焰离子化检测器
HCA	Hierarchical clustering analysis	层次聚类分析
HDL-C	High-density lipoprotein cholesterol	高密度脂蛋白胆固醇
HMG-CoA	3-hydroxy-3-methylglutaryl coenzyme A	3-羟-3 甲基戊二酰辅酶 A
IBD	Inflammatory bowel disease	炎症性肠病
IFN-γ	Interferon-γ	γ 干扰素

续表

缩略语	英文名	中文名
GC	Gas chromatography	气相色谱
GCA	Glycocholic acid	甘氨胆酸
GLP-1	Glucagon-like peptide-1	胰高血糖素样肽1
GLUT4	Glucose transporter-4	葡萄糖转运蛋白4
GSH-Px	Glutathione peroxidase	谷胱甘肽过氧化酶
LDL-C	Low-density lipoprotein cholesterol	低密度脂蛋白胆固醇
LDL receptor	Low density lipoprotein receptor	低密度脂蛋白胆固醇受体
IL-6	Interleukin-6	白细胞介素-6
LPS	Lipopolysaccharide	脂多糖
MDA	Malondialdehyde	丙二醛
MUFA	Monounsaturated fatty acid	单不饱和脂肪酸
NAFLD	Non-alcoholic fatty liver disease	非酒精性脂肪肝病
PBS	Phosphate buffered saline	磷酸盐缓冲液
PCA	Principal component analysis	主成分分析
PGC-1α	PPARγ coactivator-1α	过氧化物酶体增殖物激活受体γ辅激活因子1α
PLS-DA	Partial least square discriminant analysis	偏最小二乘法
PPAR	Peroxisome proliferator-activated receptor	过氧化物酶体增殖因子活化受体
PUFA	Polyunsaturated fatty acid	多不饱和脂肪酸
PYY	Peptide tyrosine-tyrosine	酪酪肽
SCD1	Coenzyme A desaturase 1	辅酶A去饱和酶1
SCFA	Short-chain fatty acid	短链脂肪酸
SFA	Saturated fatty acid	饱和脂肪酸
SOD	Superoxide dismutase	超氧化物歧化酶

续表

缩略语	英文名	中文名
SREBP-2	Sterol regulatory element-binding protein-2	甾醇调控元件结合蛋白2
STZ	Streptozotocin	链脲佐菌素
T-AOC	Total anti-oxidant capacity	总抗氧化能力
TBil	Total bilirubin	总胆红素
TC	Total cholesterol	总胆固醇
TCA	Taurocholic acid	牛磺胆酸
TG	Triacylglycerols	甘油三酯
TNBS	2,4,6-trinitrobenzenesulfonic acid	2,4,6-三硝基苯磺酸
TNF-α	Tumor necrosis factor-α	肿瘤坏死因子α
UPLC-Q-TOF/MS	Ultraperformance liquid chromatography quadrupole time-of-flight mass spectrometry	超高效液相色谱串联四级杆飞行时间质谱